Seed to Civilization

NEW EDITION

SEED TO CIVILIZATION

The Story of Food

CHARLES B. HEISER, JR.

HARVARD UNIVERSITY PRESS
CAMBRIDGE, MASSACHUSETTS
LONDON, ENGLAND
1990

This book is printed on acid-free paper, and its binding
materials have been chosen for strength and durability.

Library of Congress Cataloging in Publication Data

Heiser, Charles Bixler, 1920–
Seed to civilization : the story of food /
Charles B. Heiser, Jr.
—new ed.
p. cm.
Includes bibliographical references.
ISBN 0-674-79681-0.—ISBN 0-674-79682-9 (pbk.)
1. Agriculture—History. 2. Food—History.
3. Food crops—History. 4. Ethnobotany.
5. Ethnozoology. I. Title.
S419.H44 1990
630.9—dc20 89-27406
CIP

Designed by Gwen Frankfeldt

TO DOROTHY
who has kept me well fed

Contents

Preface to the 1990 Edition ix

1 In the Beginning *1*

2 Seeds, Sex, and Sacrifice *14*

3 Eating to Live *27*

4 Meat: The Luxury Food *34*

5 Grasses: The Staff of Life *61*

6 Sugar *111*

7 Legumes: The Meat of the Poor *117*

8 The Starchy Staples *134*

9 Coconut: The Tree of Life *159*

10 Oils: Sunflower and Cotton *168*

11 To Complete the Meal *177*

12 Multiplier of the Harvest *197*

13 Let Them Eat Cake? *208*

Bibliography *217*

Index *223*

Preface to the 1990 Edition

This is not a book about hunger. Rather it concerns mostly the plants and animals that stand between us and starvation. The subject can be called ethnobiology, the study of plants and animals in relation to humans. Ecology, the study of organisms in relation to their environment, is another of our concerns. In this case we are the organisms and the part of the environment of interest to us is the plants and animals that provide our food.

In this book I begin with some consideration of the origin of agriculture and why plants and animals were domesticated. The bulk of the book is concerned with basic food plants and animals, and covers where and when they were first domesticated as well as why and how they are used. I have, however, not hesitated to stray from the principal subjects from time to time when I have felt that the digression would be of general interest to my readers. There is, for example, some mention of the uses of plants and animals for purposes other than food. In this edition I have extensively revised several chapters, completely rewritten the last two, and increased the number of plants treated. In addition, I have incorporated some new photographs.

Only one chapter is given to the discussion of animals and its focus is on those most important as food. As I am a botanist, some may think that I have neglected animals in favor of plants, but in defense I can point out that we get all of our carbohydrates and nearly three-fourths of our protein from plant sources. Moreover, nearly all of the food we get from animals is in turn derived from plants. After all, life depends on photosynthesis; chlorophyll has been referred to as the green blood of the earth. The last chapter concerns current and future food problems and, perhaps, some controversial topics.

The book has been written with the general reader in mind and no particular background in biology should be necessary for understanding most of the topics. I had once assumed that the readers of a book such as this would have an elementary knowledge of human nutrition, but, judging from recent news releases, that assumption was unjustifiable, for malnutrition is not confined to the poor and uneducated but extends to the affluent and "educated" as well. Therefore, a brief treatment of nutrition is given (Chapter 3).

Although I have not tried to include all of my sources, there is a fairly extensive and updated bibliography at the end of the book. This is included primarily for those readers who would like to pursue any subject in greater detail. Although I expect the same plants and animals to continue to serve as our principal foods for a long time to come, our detailed knowledge concerning them will change as research makes more information available. Perhaps this is nowhere more true than in the realm of prehistory, for the next archaeological investigation may uncover new information regarding the "invention" of agriculture and the earliest domesticated plants and animals.

There are a number of individuals to whom I am indebted for their advice. My thanks to all of them, particularly Gregory Anderson, Virginia Flack, Gunder Hefta, Jorge Soria, and my wife, Dorothy. Thanks also to those who supplied the illustrations.

Seed to Civilization

CHAPTER ONE

In the Beginning

In the sweat of thy face thou shalt eat bread.
GENESIS 3:19

People have been on earth for some two million years and over one hundred thousand years ago essentially modern humans appeared. Until ten thousand years ago all people were hunters of animals and gatherers of plants, dependent upon nature for their food. They must, at many times during their long history as hunter-gatherers, have enjoyed full stomachs, when vegetable foods were abundant or ample game was available. Early humans certainly must have experimented with nearly all of the plant resources, thus becoming experts on which ones were good to eat. They became excellent hunters and fishers. Contrary to earlier opinion, recent studies suggest that they didn't always have to search continually just to find enough to eat and, at times, must have had considerable leisure.

There undoubtedly were times and places, however, in which people did have to spend most of their waking hours searching for food, and hunger probably was common throughout much of the preagricultural period. Certainly there could never have been much of an opportunity for large populations to have built up, even among the successful hunter-gatherers. People probably lived in small groups, for with few exceptions a given area would provide enough food for only a few. Disease and malnutrition probably contributed to keeping populations small, and it is likely that there were also some sorts of intentional population control, such as infanticide.

Then, about ten thousand years ago, food-procuring habits began to change, and in the course of time our ancestors became food producers rather than hunter-gatherers. At first they had to supplement the food they produced with food they obtained by hunting and gathering, but gradually they became less dependent on wild food sources as their domesticated plants and animals were increased in number and improved. The cultivation of plants and the keeping of animals probably required no less effort than did hunting and gathering, but in time they gave a more dependable source of food. Having a dependable food source made it possible for larger numbers of people to live together. More mouths to feed were no longer disastrous, but rather were advantageous, for with more bodies to till and reap, food could be produced more efficiently. Although some urban centers may have developed before agriculture, food production was probably the chief stimulus for the growth of villages and eventually of cities. And with the latter came civilization.

When food production became more efficient, there was time to develop the arts and sciences. Some hunter-gatherers, as was already pointed out, must have had considerable leisure, but they never made any notable advances toward civilization. An important difference between hunter-gatherers and farmers is that the former are usually nomadic whereas the latter are sedentary. But even those preagricultural people, such as certain fishermen, who had fairly stationary living sites did not develop in civilizing ways comparable to those of farmers. Agriculture probably required a far greater discipline than did any form of food collecting. Seeds had to be planted at certain seasons, some protection had to be given to the growing plants and animals, harvests had to be reaped, stored, and divided. Thus, we might argue that it was neither leisure time nor a sendentary existence but the more rigorous demands associated with an agricultural way of life that led to great cultural changes. It has been suggested, for example, that writing may have come into existence because records were needed by agricultural administrators. Plants and animals were being changed to suit needs; living in a new relation with plants and animals was, in turn, changing the way of life.

In recent years archaeological work has greatly increased our knowledge of the beginnings of agriculture, and without doubt future archaeological work will add a great deal more information. In contrast to previous generations of archaeologists who were mostly concerned with spectacular finds—tombs and temples, the contents of which would make showy museum exhibits—recent archaeologists have taken a greater interest in how people lived, what they ate, and how they managed their environment. A few charred seeds or broken bones may appear rather insignificant in a museum, but they can reveal a great deal about early human activities. As a result of recent work in archaeology, done in cooperation with scientists from many other fields, we are beginning to understand the ecology of prehistoric people in many different parts of the world.

Our knowledge of what humans ate and did thousands of years ago comes from the remains of plants and animals recovered from archaeological excavations. Unlike many tools, which were made of stone and are indestructible, foods are perishable and are preserved only where conditions are ideal. The best sites are in dry regions, often in caves, and from such sites we obtain remains to use in the reconstruction of our ancestors' diet. Other human artifacts, such as flint sickles and stone querns, or grinding wheels, may also provide clues about diet, but they leave us to speculate about what plants were being harvested and prepared, and whether these were wild or cultivated. Obviously, the record of what prehistoric people ate is very incomplete, and for many areas of the earth significant remains have yet to be found.

Drawings of animals, particularly from the later prehistoric periods, have come down to us and sometimes (but not always, by any means) can be fairly readily identified, but it is animal bones, or even fragments of them, that provide the best clues about the animals that were closely connected with people. An expert zoologist can identify species from bones, but it is not always possible to say whether remains are from domestic or wild animals.

Plant remains comprise a variety of forms. Most are seeds or fruits, but other parts such as flower bracts, stalks, and leaves are sometimes found. A few remarkably well-preserved seeds are recovered, looking as if they had been harvested only a year before, but most seeds are charred and broken. A skilled botanist can identify such plant remains, and it can often be determined if they are from domesticated or wild plants.

Another source of information about the ancient diet is coprolites—fossil feces. By suitable preparation they can be restored to an almost fresh condition (sometimes, it is said, including the odor). Whole seeds have been found in coprolites, but most of the food material is highly fragmented and requires lengthy, painstaking analysis for identification. Such analysis is highly significant because it tells us what was actually eaten, in what combinations, and whether it was cooked or raw.

Unfortunately, material collected at an archaeological dig is sometimes not accurately identified, as has been shown for some of the early archaeological reports from Peru. Fortunately, however, the material recovered from archaeological sites is usually preserved in museums and future investigators can examine the material to verify or correct identifications.

With the development of radiocarbon methods of dating it became possible to date, fairly accurately, the beginnings of plant cultivation. Sometimes radiocarbon dates, for one reason or another, may be open to suspicion, but when different materials from the same site are analyzed and several dates agree, we have fair assurance that they are correct within a few hundred years.

The evidence that has accumulated over the past several years indi-

A

B

Figure 1–1   A. Archaeological dig at Coxcatlan cave, one of the sites in the Tehuacan Valley, Mexico, that has revealed early evidence of maize. (From D. S. Byers, ed., *The Prehistory of the Tehuacan Valley*, vol. I. Austin: University of Texas Press, 1967. Used by

A

B

Figure 1–2   *A*. Einkorn wheat from archaeological site of Nea Nikomedeia, Greece. (Courtesy of W. van Zeist.) *B*. Emmer wheat from archaeological site of Nea Nikomedeia, Greece. (Courtesy of W. van Zeist.)

permission.) *B*. Increase in size of maize between c. 5000 B.C. and c. A.D. 1500 at Tehuacan. The oldest cob is slightly less than one inch long. (From D. S. Byers, ed., *The Prehistory of the Tehuacan Valley*, vol. I. Austin: University of Texas Press, 1967. Used by permission.)

cates that agriculture probably had its origins in the Near East*—
although not necessarily, as earlier supposed, in the fertile river valleys
of Mesopotamia (which were to be important centers of early civilization), but more likely in the semiarid mountainous areas nearby. Dates
determined for flint sickles and grinding stones discovered in these
areas indicate that before 8000 B.C. humans had likely become collectors
of wild grain, and there is evidence that a thousand or so years later they
were actually cultivating grains and keeping domesticated animals. Several sites are now known in the Near East (see Figure 1–3) that give
evidence of early agriculture. One of the first sites to give such evidence
was at Jarmo, in Iraq, where investigations were conducted under the
direction of R. J. Braidwood. In deposits dated at 6750 B.C. seeds of
wheat and barley and bones of goats were found. Other evidence of
cultivation, dating from approximately the same time, has been found at
several other sites in the Near East. Since the plants there apparently
represent cultivated species, we must suppose that there was an earlier
period of their incipient domestication, which may have lasted for a few
hundred years or more. How long it takes a plant to become fully domesticated cannot be answered precisely and it probably varies considerably from species to species. In deposits accumulated after 6500
B.C. we find evidence of other plants being cultivated in the Near East
and Greece, and bones of various domesticated animals become more
abundant.

Other centers of agriculture developed in the Old World. Whether
these developments were stimulated by knowledge of agriculture in the
Near East or whether they were independent developments is not certain, but the fact that some of them were based on completely different
plants from those of the Near East might support the latter view. For a
long time southeastern Asia has been considered an ancient center for
domesticated plants, but until recently there was no archaeological support for this view; this is not wholly unexpected, since the climate for the
most part in that area of the world is hardly conducive to preservation of
food remains. In 1969, however, a report was published of an assem-
blage of plants from Thailand, including possibly a pea and a bean,
dated at 7000 B.C. As it is not definitely clear whether the recovered
plants represented wild or cultivated species, we cannot yet say that
agriculture was practiced as early here as it was in the Near East. Rice,
which was to become the basic food plant of southeastern Asia, was not

*The *Near East* (see map, Figure 1.3) is the term used by archaeologists to refer to the
countries of southwest Asia. The term *Middle East*, widely used in the news today, includes the Near East.

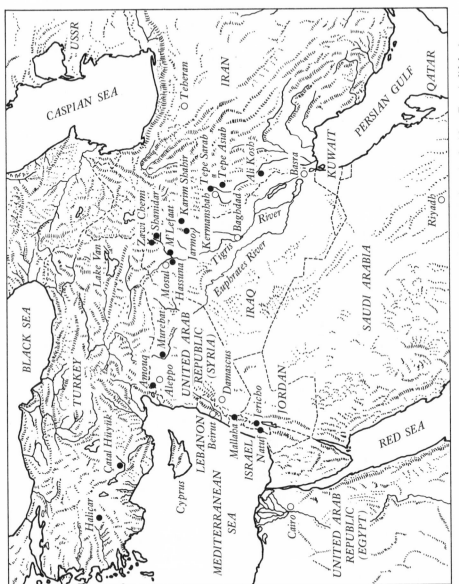

Figure 1–3   Selected archaeological sites that show evidence of early agriculture in the Near East (*solid dots*).

brought under cultivation until 5000 B.C., considerably later than the cereals in the Near East. There is some indirect evidence that agriculture may have been practiced in New Guinea at 7000 B.C. but no actual remains of plants have been found.

In the New World, agriculture had its origins in Peru and Mexico. Although a domesticated bean from Peru has been reported as ten thousand years old, agriculture apparently did not begin in the Americas until a few thousand years later than in the Near East. Through a series of excavations directed by R. S. MacNeish, we now have a remarkable sequence of plants giving evidence of the period of incipient domestication in Mexico. Some indication of the earliest cultivated plants is found in the mesquite-desert regions of southwestern Tamaulipas, with gourds, squashes, beans, and chili peppers being found at levels dated at between 7000 and 5500 B.C.

Following investigations at Tamaulipas, MacNeish made deliberate efforts to search for evidence of the domestication of maize, which eventually became the most important plant in the Americas. A group of caves in the arid highlands near Tehuacan in south central Mexico showed promise, and a series of excavations was begun in 1961. The results give us the best picture yet of the transitional stages leading to full-scale agriculture. Humans were probably in the Tehuacan area by 10,000 B.C. and for several millennia they depended on wild food sources, both plant and animal. Gradually more and more plants were cultivated, some perhaps having been domesticated at this site, others introduced from other regions. The first suggestion of cultivated plants occurs in material dated at about 5000 B.C., with maize, squash, chili pepper, avocado, and amaranth being found. These plants were definitely cultivated during the next period (4900–3500 B.C.), together with various fruits and beans toward the end of the period. During the next thousand years other plants were added, including cotton and two new kinds of beans. The dog, which is known from historic records to have been an important food item in Mexico, is first associated with humans in the archaeological record at this time. At about the beginning of the Christian era, the inhabitants of Tehuacan had also acquired the turkey. From remains of the same period there are reports of some other plants: guava, pineapple, and peanut. The presence of these plants would be of particular interest, for the peanut is definitely South American in origin and the pineapple and guava perhaps are also, which would suggest that the peoples of this area had contact with South America at this time. None of these plants has been found in any other archaeological sites in Mexico to date. A study of historical records of both peanuts and pineapple would suggest that they arrived in Mexico recently, perhaps after the coming of the Spanish.

Figure 1–4 Selected New World archaeological sites that show evidence of early agriculture (*solid dots*).

Another early development of agriculture in the Americas occurred in Peru, perhaps even earlier than in Mexico. Two kinds of cultivated beans and a chili pepper, dated at around 6000 B.C. or earlier, have been recovered in a highland valley at Callejon de Huaylas in north central Peru. Previous to this discovery, archaeological plant material had been found in dry coastal sites, such as Huaca Prieta in northern Peru. Gourds, squashes, cotton, lima beans, and chili peppers are among the first plants cultivated on the coast; present evidence indicates that agriculture developed here about two thousand years later than in the highlands.

At present it is difficult to say whether agriculture in the Americas appeared first in Peru or Mexico. The fact that many of the same plants were cultivated in the two areas might suggest that agriculture spread from one of the areas to the other, but the chili peppers and the squashes of the two regions belong to different species, and it seems likely that the common bean was domesticated independently in Mexico and Peru. Thus, although the possibility remains that agriculture, or at least the idea of growing plants, diffused from one area to the other, it is just as likely that agriculture arose independently in Mexico and Peru. That there was diffusion between the two areas at a later time is clear because maize, which almost certainly had its origin in Mexico, appears in Peru by 2500–3000 B.C. and, as already mentioned, certain South American plants may have appeared in Mexico in pre-Columbian times.

From the foregoing account it can be seen that agriculture arose in widely separated parts of the earth, probably quite independently from place to place. But agriculture began in the Old World more than a thousand years earlier than it did in the New—could the idea of agriculture have come to the New World from the Old? The New World was peopled by immigration across the Bering Strait long before agriculture was known, and if there were subsequent crossings, it was by hunters rather than agriculturists. Thus, we would have to postulate a long ocean voyage at a very early date to account for agricultural knowledge being brought to the New World. Some anthropologists have postulated that there were such voyages in prehistoric times, but much later than the time at which agriculture was established in Peru and Mexico. Therefore, it seems highly unlikely that agriculture had but a single origin. It is, in fact, likely that it had several origins in both the Old World and the New, although some people still believe that it was "invented" only once.

Agriculture diffused to other parts of the world from these early centers. From the Near East it spread to both Europe and Africa. In the New World it has generally been thought that agriculture went from Mexico to eastern North America. Recent evidence indicates that some native plants—the sunflower, sumpweed, and a chenopod—were in cultivation in eastern North America long before the arrival of beans, maize, and tobacco from Mexico. Squash was also an early cultivated plant in eastern North America, but it appears that it was domesticated independently in Mexico and eastern North America. Thus it now seems likely that plant cultivation also had a separate origin in what is now the United States.

An examination of the list of food plants from all the early sites in both the New and the Old World reveals that all of the plants were propagated by seed. A large number of present-day food plants, including such important ones as the white and the sweet potato, manioc, yams, bananas, and sugar cane, are propagated vegetatively—by stem cuttings, tubers, or roots—rather than from seed. Some people, notably the

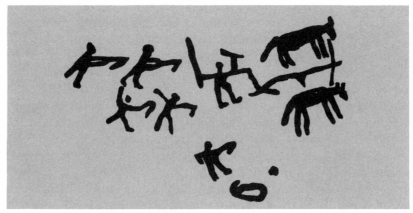

A

B

C

Figure 1–5 A. Hoeing and horse-drawn plow, reproduced from cave drawings in northern Italy. (Redrawn from a photograph by Emmanuel Anati.) B. Hoeing and ox-drawn plow, from decorations on a tomb at Beni Hasan, Egypt. (Courtesy of Egypt Exploration Society.) C. Present-day plowing with oxen in Egypt. (Courtesy of FAO).

geographer Carl Sauer, have reasoned that cultivation of plants probably began with vegetative propagation, arguing that such cultivation is much simpler than seed planting. There is also some evidence from Old World mythology suggesting that vegetative cultivation is older than seed planting. The archaeological record, unfortunately, has not been able to provide us with clear-cut answers, for many of the vegetatively cultivated plants are crops of the wet tropics, areas where the preservation of prehistoric food materials is rather unlikely. Moreover, even in dry areas, tubers and other fleshy plant parts are far less likely to be preserved than are relatively dry materials, such as seeds. While we cannot, perhaps, entirely rule out the possibility that agriculture based on vegetative propagation was earlier than seed-propagation agriculture, it seems fairly clear that it was seed planting that led to the most profound changes in our way of life. All the early high civilizations whose diets are known to us were based on seed-reproducing plants—wheat, maize, or rice—with or without accompanying animal husbandry.

Following the domestication of plants and animals, the next great advance in agriculture came with the control of water. Irrigation arose in the Near East around 5000 B.C. and in Mexico shortly after 1000 B.C. With irrigation, considerably more food could be produced in many areas; as a result, a few people could produce enough food to feed a large population, permitting others to spend their time in pursuit of the arts and crafts and of religion. Elaborate temples, many of them standing today, were constructed by early societies that had perfected methods of irrigating their crops and testify to the amount of human labor that was made available for other occupations.

Another important development in Old World agriculture was the use of animals to prepare the fields for planting, which was never done in the New World in prehistoric times. Along with this difference there was a basic difference in planting techniques. In the Old World the cereals (wheat, for example) were planted by broadcasting handfuls of grain, whereas in the New World the grains of maize were planted individually.

With the domestication of plants and animals there should have been a dependable food supply and, so it might be thought, hunger should have disappeared from the earth. As any intelligent person is acutely aware, however, hunger is still very much with us today. Harmony with nature has yet to be established. With the advent of agriculture, humans began changing their environment drastically. Irrigation, which initially led to greater food production, eventually destroyed some of the most fertile areas. Without adequate drainage, irrigation leads to an accumulation of salts in the soil that few plants can tolerate. That this happened in prehistoric times in the Near East is evident from archaeological findings; for barley, which is more salt tolerant than wheat, replaced the latter plant in some regions after irrigation was developed. The use of animals to till the soil led to increased areas being planted, which in time must have been accompanied by increased soil erosion. Then, along with the plants and animals that people brought under their control came others that they did not want and could not control. Rusts, smuts, and weeds soon found cultivated plants and fields a fertile territory for their development, and insects, rodents, and birds moved in to appropriate the new foods for themselves. Competition for the more fertile agricultural land led to warfare on an escalating scale, for which the powers of some of the domesticated animals were used. Hunger has always accompanied war.

Deserts now occupy many of the areas where high civilizations once flourished. Natural climatic change may in part be responsible for some of these deserts, but humankind most likely contributed through misuse of soil and water. Alteration of the environment, which began in a modest way ten thousand years ago, continues in the present on a scale never known before, and concern for the future of agriculture is warranted.

CHAPTER TWO

Seeds, Sex, and Sacrifice

O goddess Earth, O all-enduring wide expanses!
Salutation to thee.
Now I am going to begin cultivation.
Be pleased, O virtuous one.

ANCIENT SANSKRIT TEXT

The work of the archaeologist has revealed a great deal concerning the "invention" of agriculture. We now have some idea about where and when it occurred and what plants and animals were involved, but we do not know *why* plants and animals were domesticated. The answer may be very simple: It was desirable to have a dependable source of food close at hand, and what would be more obvious than to bring animals into confinement and to grow plants in some suitable place near the home? In James Michener's best-selling novel of a few years ago, *The Source*, we read that thousands of years ago the wife of Ur transplanted wild grain near her dwelling; although her first efforts failed, she eventually succeeded in cultivating it. This fictional account is of interest for two reasons. First, it postulates that planting began with woman. It seems quite possible that women deserve the credit because they were probably responsible for the gathering of seeds and roots and the preparing of meals and would therefore have had a much more intimate knowledge of plants than did men. Second, it serves to illustrate the "genius theory" of the origin of agriculture, which would have it that agriculture arose through the efforts of a single brilliant person. Most archaeologists have been unwilling to accept such a hypothesis, perhaps because it explains nothing of the circumstances leading to cultivation. This is not to say that people were any less intelligent or observant ten thousand years ago than they are today; that a seed germinates to give rise to another plant of the same kind was probably well known to those who depended on seeds as their main source of food. Knowledge of seed germination would not necessarily have led, however, to the planting of seeds. As Kent Flannery has pointed out, "A very basic problem in human culture is why cultures change their mode of subsistence at all." Our problem is to explain the change from hunting-gathering to farming. The archaeological record does not provide any definite answers.

Humans were not the world's first farmers; ants grow plants (fungi) and tend animals (aphids), and probably did so long before humankind appeared on the scene. No one has suggested, however, that people acquired their knowledge of agriculture from ants. In the past it was thought that humans naturally tend to improve themselves and that it was therefore only natural that they would turn to agriculture. Until fairly recently there has been little serious inquiry into how agriculture actually began. One of the first serious considerations of the origins of agriculture was that of V. Gordon Childe, who postulated in 1936 that a climatic change resulting in desiccation in the Near East brought people and animals together where there was water and that this association stimulated the domestication of animals. It was Childe who proposed the term *Neolithic Revolution* for the initiation of agriculture.

A widely cited hypothesis is that of Carl O. Sauer, who in 1952 advanced the idea that agriculture arose among fishermen in southeastern Asia. Fishermen, he reasoned, had a dependable food source nearby and would therefore have been more or less sedentary, giving them the time and stability to experiment with plant cultivation. He believed that, as far as agriculture was concerned, necessity was not the mother of invention. Also still often cited is the "dumpheap" hypothesis of Edgar Anderson from the same year. In the waste places that people created near their habitations, he visualized a breeding ground for weeds, some of which would have been candidates for the first domesticated plants.

One of the first modern studies of the origin of agriculture was that of R. J. Braidwood, who with a team of workers, including biologists, carried out archaeological excavations in Iraq beginning over a quarter of a century ago. As a result of this study, Braidwood argued that there had been no major climatic change in the Near East and that food production developed "as the culmination of ever increasing cultural differentiation and specialization of human communities." He assumed that agriculture would naturally accompany familiarity with plant and animal resources.

Childe's theory of climatic change remained unsupported by evidence for some time. Some archaeologists maintained that the climate in the Near East had been relatively stable since long before agriculture began. Recently, however, evidence has been supplied that there was indeed some change in climate in the Near East, though not of the nature Childe had supposed.

In 1968 an analysis by H. E. Wright, Jr., of pollen deposits in two lake beds in Iran indicated that there was a shift in climate about eleven thousand years ago. Through the analysis of fossil pollen it is possible to determine what plants grew in a given area in former times, and from this we may infer the climatic conditions under which the plants grew. The pollen analysis in Iran indicates a shift from a cover of herbaceous plants to one in which oak and pistachio trees predominated; this is interpreted as a change from a cool steppe to a warmer and perhaps moister savanna. Such a change would obviously have had an effect on the plants and animals present and consequently on food-procuring habits.

How else could the environment have figured in the origin of agriculture? In both the Near East and Mexico the earliest known sites of agriculture are from semiarid, somewhat hilly or mountainous country. Is this simply due to a natural bias for preservation in such settings? Evidence of agriculture is certainly far more likely to be preserved in arid or semiarid regions than in moister areas. Or is it possible that such places did favor the development of agriculture? Although this sort of environment may not seem ideal, there would have been certain advantages. These areas of diversified terrain would have provided a number of microenvironments appropriate for different species of plants and animals, which would in turn have offered a considerable array of wild foods and potential domesticates. Only limited travel would have been necessary to secure enough to eat. Thus communities could have been sedentary at least at certain times of the year, a necessity for the establishment of cultivation. Low rainfall, as long as it was adequate for plants at certain seasons, may have offered many advantages to early agriculturalists. There would have been no heavy plant cover to remove before planting and there would likely have been fewer weeds and plant pests, both insect and fungal, than in better-watered areas. While it seems fairly evident that environment, and perhaps even climatic change, may have played a role in the origin of agriculture, it is difficult to see how these could be the sole factors or even the most important ones.

In a most ambitious attempt to solve the problem of the origin of agriculture, Charles A. Reed assembled a group of experts from around the world in 1973. No consensus emerged, but the 1013-page volume resulting from the conference provides a number of provocative and stimulating articles on the subject. One of the participants in Reed's conference, Mark Cohen, later wrote his own book on the subject in which he developed the following thesis: The only factor that can account for the irreversible and nearly uniform emergence of agriculture throughout the world is the growth of populations beyond the size that hunting and gathering would support. The events leading to the development of agriculture in various parts of the world, Cohen maintained, show a remarkable parallelism. Over eleven thousand years ago hunters and gatherers had occupied all the lands that would support their life styles and they were forced to turn more and more to unpalatable foods. The people who started agriculture were not verging upon starvation; the population pressure was "nothing more than an imbalance between a population, its choices of food, and its work standards, which forced the population to change its eating habits or to work harder." Although agriculture did not provide a better diet, greater dietary reliability, nor greater ease in the quest for food, it did provide more calories per unit of time and per unit of space than could hunting and gathering.

Cohen is not the first to maintain that demographic pressure was responsible for the origin of agriculture, but he is the first to develop the thesis at such length. Others, including some of the participants in Reed's conference, had denied that population pressure figured in the origin of agriculture. There is also some doubt that population pressure had developed in all of the areas where Cohen thinks agriculture began—he assumes that it had four or more origins. It is difficult to imagine that the small returns from the first attempts at cultivation would have been of much profit to people who were beginning to experience food shortages. Would they have been willing to save some seeds for planting rather than eat them all? Cohen supposes that the concept of agriculture is simple and was widely known among primitive peoples and that all that was necessary to implement it was population pressure. But the sowing of seed and even cultivation (although not always for the sake of food rewards) must be more ancient than the archaeological record has thus far revealed. Agriculture could have been implemented in different places for a variety of reasons. Perhaps population pressure was responsible for the acceptance of agriculture in some places, but it was not necessarily the primary factor everywhere.

In a recent book David Rindos has proposed a radically different hypothesis to account for the origin of agriculture, maintaining that it developed unconsciously on the part of people. He sees no difference in the domestication of fungi by certain ants, for example, and the domestication of plants by people. Darwin in addition to natural selection recognized two types of human selection—methodical and unconscious. Rindos calls upon the latter to have brought about the changes in plants that led to domestication. Thus there was a coevolution of the plants and people, but no intentionality was involved on the part of the people. One may accept that a coevolution of a sort developed and that unconscious selection played a major role in the process of domestication, but Rindos overlooks one important aspect. For unconscious selection to have produced the changes in the plants, people would have had to plant seeds first, an intentional act. Those plants whose seeds were not deliberately planted and that coevolved with people became weeds.

In 1986 Kent Flannery, who has the advantage of having worked at several sites of early agriculture in both the Old World and the New, pointed out that one should not look for a single cause of the origin of agriculture; it is to be explained instead by the interaction of a number of factors. "The origin of agriculture," he writes, "involved both human intentionality and a set of underlying ecological and evolutionary principles." Drawing insights from some of the recent hypotheses discussed here and in the works of L. R. Binford and F. Hassan, he proposes a multivariate model for the origin of agriculture in Oaxaca, Mexico.

With the exception of Erich Isaac, who identified the origin of vegetative cultivation with the cutting up and burying of plant parts in a ritual enactment of the primeval killing and burial of a god, few modern scientists have postulated a religious motive for the origin of agriculture. Certainly both plants and animals have had an intimate association with religion in early historical times. In order to appreciate the possible roles of religion in the origin of agriculture, we must make an excursion into people's beliefs concerning themselves and their environment in prehistoric times.

Our knowledge of early religion is, of course, very fragmentary. Interpretations of the religious significance of material remains are highly speculative. It seems quite evident, however, that birth, death, and food were of fundamental importance to early humans and that concern with these affected most of their activities. As there were many mysteries connected with all of these facts of life, they became interrelated in human thought. In Paleolithic times animals, although they were killed for food, were nevertheless considered akin to humans. The seasonal cycles of the death and rebirth of vegetation were thought to be related to the human life cycle. The worship of trees and plants was probably an early manifestation of religion and was carried down through the Neolithic period to become an important part of formal religion in early historical times. Such beliefs still survive in the folklore of many peoples. In trees and other vegetation human beings recognized a life-giving power akin to their own and to that of animals. Human fertility, that of animals, and that of vegetation were obviously considered to be closely related.

Associated with human burials from early times in many parts of Europe and Asia are small female figurines, or Venuses, made of bone, stone, or ivory. The face may be entirely lacking or crudely represented

Figure 2–1   Mother-goddess figurines from the Near East and Europe.

and the sexual features are often exaggerated. The well-developed abdomen on some figures is thought to represent pregnancy; figures with a prominent vulva perhaps indicate women giving birth. The rather extensive distribution of such figurines has been interpreted as evidence of a widespread mother-goddess fertility cult in Neolithic times; eventually various distinct goddesses were worshiped as regional religions developed.

Recognition of the sexual significance of the male* apparently increased during later Neolithic times, for phallic symbols from this period have been found in various regions by archaeologists. Eventually a young god—in some places considered to be a brother or son of the goddess—became a characteristic of early religions. The union of the goddess with this god was then regarded as being responsible for fertil-

---

*That the female had a role in fertility, of course, could never be doubted. Exactly when the full significance of the male was recognized is not known. Some present-day primitive groups do not recognize the male's contribution to procreation, believing rather that intercourse may be necessary to allow a spirit to enter the womb or to make childbirth easier. This idea contrasts strongly with that prevalent among Western peoples during the late Middle Ages, when it was thought that the sperm contained the fully made child (the *homunculus*) and the female simply served as a house for its early development.

ity. Among some human groups, the living king was thought to be a god and as such was an important figure in sacramental marriage ceremonies relating to fertility. From this it is not difficult to see how all human sexual intercourse became symbolically associated with the fertility of plants and animals. In later times, human sexual intercourse became a part of festivals held in the fields at the time of planting to promote growth of the crops and of other festivals associated with agriculture. The plow itself seems to have been first designed as a phallic symbol, representing man's role in bringing fertility to mother earth. Sexual offenses were thought to impair this fertility. Thus we see that in early times sex was considered to be sacred among some peoples.

With the development of agriculture, other gods joined the mother goddess and her consort to share their duties; the earth, the sky, the rain, and other natural elements were among the special domains of the various gods. In many cultures the principal male god of the pantheon (Zeus is one of the best known), who had been born of the mother goddess, eventually assumed a dominant role. The relationships among the gods became complicated and their responsibilities less clear cut, but there were always some whose main concern was fertility, being remnants of the earlier fertility cults. Fertility ceremonies continued during the Roman Empire and the excesses committed at some of the festivals led the Romans to decree laws against their observance. Such ceremonies are still widely practiced among "primitive" people in many parts of the world today and sometimes by "advanced" people as well,* although not always in a form recognized as such. Most people participating in May Day celebrations today are probably unaware that the original purpose of the festival was to promote the well-being of vegetation. Other remnants of ancient practices may still be found in the United States; some farmers, for example, believe that planting should be done during certain phases of the moon and that if a menstruating woman walks through a garden the crops will fail.

The temple prostitutes mentioned in the Bible were participants in rituals derived from the earlier sacramental marriages. The prostitutes and the fertility festivals were denounced by the Israelites, who originally were pastoral desert-dwellers and not tillers of the fields. As their religion was monotheistic and their god, Yahweh (or Jehovah), was not a fertility deity, they could not view the ceremonies as legitimate religious activity or the women as anything but harlots. It has recently been

---

*Thomas Tryon's *Harvest Home,* a fictional account of the people of a modern New England village who still practice the ancient planting and harvest rites, became a best-seller in 1973 and was later made into a movie for television starring Bette Davis.

pointed out that many of our present-day environmental problems may stem from the Judeo-Christian concept that the earth and everything on it were put here solely for our use. Would things perhaps have been different if the religions of the developed nations had derived from the fertility cults, with emphasis on reverence for mother earth and her creatures, both plant and animal?

How early sacrifice developed is not known, but there is evidence that it was practiced at Tehuacan before 5000 B.C. and became well established in most early agricultural cultures. Some people have postulated that humans were the first victims and were later replaced by animals, but it is perhaps just as likely that humans replaced animals as cultures reached a certain stage of advancement. Practically all of the domesticated animals have been used in sacrifice at one time or another, sheep, goats, and cattle all being prominent, as readers of the Old Testament are aware. No animals were more important than cattle. A cattle cult apparently was well established at Çatal Hüyük in Turkey in about 6000 B.C. As part of fertility cults, cattle became associated with the gods themselves and became prominent figures in many primitive religions.

Various reasons have been advanced to explain sacrifice, the simplest being that it was to honor or appease the gods. It is likely that it was much more complicated than that. Sacrifice probably had a dual role.

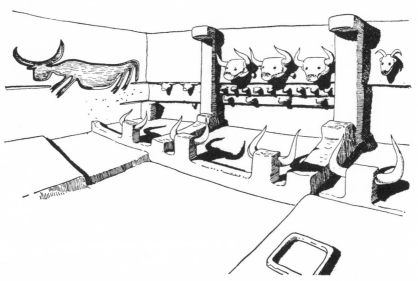

Figure 2–2 Restored cattle shrine at Çatal Hüyük, c. 6000 B.C. (From James Mellaart, *Çatal Hüyük: A Neolithic Town in Anatolia.* London: Thames and Hudson, Limited, 1967. Used by permission.)

Figure 2–3
Inca sacrificing a llama. (From Poma de Ayala, c. A.D. 1600.)

The human or animal being sacrificed represented the grain or the produce of the field and, at the same time, the people who were to partake of it. The sacrifice would bring about a desanctification of the plant to make it safe for humans to eat; it would also assure a future bountiful harvest of the fields. The victim in some cultures was the king. As this didn't prove to be popular with kings, in later times a lesser person was substituted. The dying king or his representative symbolized the dying vegetation. He was then replaced by a new king to represent the resurrection of vegetation and of life itself. After the abandonment of human sacrifice, effigies were sometimes used to serve the same purpose, a practice that survived until recent times in some places. That human sacrifice may not have completely disappeared, however, is evident in an account from Tanzania. In 1959 several farmers of the Wangi tribe were arrested for violation of a witchcraft ordinance. The farmers were suspected of *Wanyambuda,* an ancient tribal fertility rite in which fields were sprinkled with "medicine" made of seeds, blood, and parts of human bodies. Animal sacrifice certainly still exists. Among the Aymaras of highland Bolivia, for example, llamas are killed in special ceremonies and their blood is sprinkled on the newly planted potatoes.

Although this account of primitive religion is oversimplified and very incomplete, there can be little doubt that early religion was intimately involved with the quest for food. If this is true, we might inquire if in religious beliefs we can find some clues about how domestication began.

The dog was the first domesticated animal; dogs have served as human food in historic times and doubtless did in prehistoric times as well. In some cultures the dog may have been more important as a hunting companion than as food, and the suggestion has been made that the dog aided greatly in domesticating other animals, just as it continues to serve pastoral people today. It seems unlikely that a religious motive could have been involved in the domestication of the dog.

It is also entirely possible that religion had nothing whatsoever to do with the domestication of other animals. The presence of dogs may have inspired attempts at other domestications. Perhaps people brought home young animals whose mothers had been killed. The young animals may have been nursed by the women and become pets, leading to their domestication. This hypothesis has been considered the most likely by some authorities on the subject. Others have postulated that through herding of wild animals, increasing control over them developed and domestication eventually occurred.

The idea that religion may have been involved, however, is not a new one. Eduard Hahn, A German geographer who published just before the turn of the century, maintained that cattle were domesticated in order to secure animals for sacrifice at lunar fertility ceremonies. Hahn believed that cattle were chosen for such sacrifice because their crescent-shaped horns resemble the new moon. Other reasons could be suggested. The wild bull was a ferocious animal and would have inspired both fear and admiration. Perhaps men sacrificed young male animals to secure their strength and virility. That cattle became highly preferred for sacrifice and eventually even became sacred in certain cultures is quite evident, but this, of course, in itself does not prove that cattle were domesticated for religious rather than utilitarian motives.

Recent archaeological discoveries indicate that cattle were not the first ruminant animals to be domesticated, both goats and sheep having been earlier. Are we to suggest that perhaps both of these animals were domesticated for religious rather than economic reasons? Certainly both sheep and goats were widely used for sacrifice, and there is an old Sumerian incantation that refers to the sacrifice of a wild goat. If wild animals were used for sacrifice, they had to be captured alive and kept until the appropriate time. This could have been the first step in domestication. It is, of course, not necessary to try to account for the domestication of all animals through religious considerations, for once one

A   B

Figure 2–4 *A.* Prize cattle decorated for a festive occasion, India. (Courtesy of FAO.)
B. "Demon" guarding field of modern rice, India. Improved strains of plants and animals
exist side by side with ancient traditions and ceremonies in many parts of the world.
(Courtesy of Rockefeller Foundation.)

animal had been domesticated the idea of domesticating others certainly
might have occurred to the same people, or to others who learned of it.

Are there ways in which religion might have influenced the first planting of seeds? Primitive cultures must have had many rituals and ceremonies associated with both planting and harvest. Though we have no
direct knowledge of these, we can make conjectures from the knowledge we have of early historical times and from primitive cultures that
still survive. Prominent among these are ceremonies devoted to the
"first fruits" or the "last sheaf" of the harvest, ceremonies that often
imply a belief in spirits of the plants. In fact, we know from early historical record that the harvest has not always been a joyous occasion, as
might be supposed, but was formerly accompanied by much sadness
and lamentation as the body of the grain spirit was reaped. As a propiti-
ation to the spirit, people might have returned a token offering of the
seeds collected, either the "first fruits" or "last sheaves." This offering
could have served the same purposes as a sacrifice, removing a taboo
from the plant to make it safe for mortals to eat and at the same time
assuring a rich growth of the grains in the following year.

The seed offering might have been scattered over the field from which it had been gathered. People who regarded the last sheaf as sacred, for example, saved it for scattering over the field along with their seed in the next season. The seed offering might actually have been buried in the soil, recognition that mother earth was the source of life. Certain Arabs are known to have buried a seed offering and marked the grave with stones.

Thus, we might postulate that the first seed planting was a magico-religious act to appease the gods. Such rituals, we would have to assume, took place among preagricultural seed collectors, and we can imagine the next steps. People somewhere would have recognized that the sacred sowing yielded plants. These would have been harvested in the next season, and some of the seeds would have again been returned to the gods via the soil. In time ceremonial plantings would become larger and larger, and intentional cultivation would be on its way. Sacred gardens, such as the gardens of Adonis in the Near East and Greece, were known in early times. Our knowledge of these comes from historical times, but one might ask if they antedate agriculture—stemming perhaps from a "first fruits" planting. Could they have originally been grown solely for religious purposes and later have served as a precursor to agriculture?

The "first fruits" and "sacred gardens" hypotheses for the origin of planting are obviously nothing more than speculation, and a number of objections can be raised to them. They do offer, however, a possible explanation for the rapid improvement of cultivated plants once planting was initiated. We might ask why people would save their best seeds for planting rather than eating them. Obviously, if the seeds were for the gods, they would have been the largest and most nearly perfect, or perhaps have been from plants showing unusual characteristics. We might postulate that artificial selection began to operate with the first offering of seeds to the spirits of the plants.

Although it seems too far-fetched to deserve much consideration, another possible explanation of the beginning of seed cultivation might be entertained: Could the origin of the planting of seeds somehow be associated with human reproduction? If the earth, as a manifestation of the mother goddess, was regarded as the womb for vegetation, perhaps there was some concept that seed would have to be planted in her, just as men plant their "seed" in women. This would mean that men would have to have had some concept of their own role in reproduction at an early date. It would hardly seem unreasonable that some people did appreciate the significance of the male in human reproduction ten thousand years ago. The historical record is too late to be of any help to us, but it may be significant that in Sumerian, the earliest written language, the word *numum* was used for both seeds of plants and the "seeds" of animals. We also find that later the Greeks used the same word for both seed and human sperm.* Some early peoples were apparently aware that a sexual union is involved in the production of fruit and seeds. From the art work on certain early Babylonian monuments, we know that people in parts of the Near East placed clusters of flowers from male date trees in the female trees in order to secure better fruit set. Although these representations were made long after the origin of agriculture, such a sophisticated knowledge must have had much earlier antecedents. This early recognition of the role of male flowers is all the more amazing when we realize that sex in plants was not "discovered" until the end of the seventeenth century and the idea was not generally accepted until much later.

On the one hand, religion may have nothing to do with the origin of agriculture. On the other, it may be a factor that should enter into the multivariate explanations of both animal and plant domestications. We may never be able to explain the origin of agriculture, but in view of its importance to the development of humankind, one should be allowed some speculation. Evidence to support such speculation is meager, but the problem is intriguing. Perhaps future discoveries will bring us closer to the truth.

*We know today, of course, that a seed and a sperm are not at all equivalent. A seed contains the embryo of a new plant; it develops from an ovule after there has been a union of a sperm with the egg contained within the ovule. The ovules, sometimes incorrectly called "immature seeds," or "unfertilized seeds," are found in the ovary of the flower of the higher seed plants. As the ovules develop into seeds, the ovary becomes a fruit.

CHAPTER THREE

Eating to Live

If you do not supply nourishment equal to the nourishment departed, life will fail in vigor; and if you take away this nourishment, life is utterly destroyed.

LEONARDO DA VINCI

Human beings are, of course, omnivorous. Some people have held that before agriculture developed, meat and fish were the principal foods, and that our ancestors turned to plants only when animals became scarce. It is not certain that people have always been meat eaters, for monkeys and apes are primarily vegetarian, and very early humans may well have been too. Nevertheless, the association of broken animal bones with some early human remains suggests that the meat-eating habit was acquired very early.

The omnivorous character of our species helps explain how we acquired such a wide distribution over the earth's surface. Humans could find suitable foods almost everywhere they went. Although as a species we eat just about everything, any particular human community selects certain plants and animals for consumption. This is certainly true today, and probably extends far back into the prehistoric period. We can surmise that early humans experimented at times with all of the possible food resources of their environment,* but some foods became preferred over others. We know little of why they made the choices they did, but it has been suggested that palatability—such things as taste, texture, odor, and color—played an important role. But did these choices assure early humans of all the necessary nutrients? John Yudkin answers in the affirmative, stating that when people ate what they liked, they ate what they needed. Obviously had people not eaten what they needed they would have been eliminated by natural selection.

*Sometimes, of course, the experiment ended in death when someone sampled too much of a poisonous plant. At other times the result may have been unexpected: certain mushrooms, for example, would have caused hallucinations. Plants causing such effects may then have been put to repeated use, either because people attached religious significance to the effects or simply because they liked them. Some plants may have been used in ways other than for food before a food use was acquired. Hemp, *Cannabis sativa*, is used for its fiber, for its edible oil, and, as marijuana, for its euphoric effect. Which of the uses was first acquired is not known.

Use of fire was one of the early important cultural traits. Fire was not only a source of warmth, a means of protection from wild animals, and a tool to assist in capturing animals but a new way to prepare food as well. Cooking has the effect of making animal protein more readily available for human use and of breaking down the starch granules of cereals so that they are more easily digested. The acquisition of fire altered eating habits. It made available foods that in the raw state were scarcely edible, or even toxic. The improved flavor that results from cooking surely must have been appreciated then as now. Use of plants for food probably increased greatly when fire came into use.

Food production also brought changes in eating habits, as well as new problems. It has been postulated that many people changed from a primarily meat diet, rich in protein, to one comprising largely cereals, which are mostly carbohydrate. If the dependence on cereals were too great, deficiency diseases could have developed, and from bones in the archaeological record we know that this indeed happened in some places. Other kinds of disease come with the development of agriculture. Storage of food would have brought rats, which can carry disease; and some diseases of domesticated animals, such as anthrax and brucellosis, can be transmitted directly to humans. People living in concentrated populations in newly developed urban areas would have been subject to epidemics.

Food problems are still very much with us today. Some of these, such as hunger, are the subject of Chapter 13. Some of the problems stem from the fact that many foods today are highly refined; others arise from the use of readily available or easily grown foods, such as manioc, that may have replaced more nutritious crops. It is no longer true that if people eat what they like, they eat what they need, for in many parts of the world today individuals may eat an abundance of what they like from the food available and not receive adequate nutrition. Life expectancy has shown a dramatic increase in the last century. While some of this is due to improved nutrition in spite of the availability of highly refined foods, more of it is to be credited to advances in medical science.

It might be thought that most people in the United States have been exposed to the rudiments of nutrition in school. Either this is not true or they paid little attention to what they were taught, for this country has been characterized as a nation of nutritional illiterates. In a 1988 report on nutrition and health, the Surgeon General of the United States warned that North Americans are eating their way to early graves. Malnutrition in the Third World is much publicized, but there is also malnutrition in the United States today. Only part of this results from poverty; some of it occurs among people who could buy the proper foods, but rely on the ridiculous propaganda of advertisements, diet books, and health-food "authorities." It has been estimated that $500 million a year is spent on food nostrums in the United States. Apparently many people fail to realize that weight is nearly always a function of how much food goes into the body and how much is burned through exercise. As a result of poor diets, health costs in the United States run into several billion dollars a year.

Primitive people lived and reproduced without any special knowledge of nutrition (neither, of course, were they influenced by advertising propaganda). Many people today also get along well without it, but others suffer—in both the developing and the developed nations—from not having or not heeding information about nutrition.

Nutrition is the science that deals with the effects of food on the body. Unfortunately, more is known about the nutritional needs of certain domesticated animals than is known for humans, the primary reason being that animals can be studied experimentally in ways that humans cannot. The nutrients are carbohydrates, lipids, proteins, vitamins, minerals, and water, the first three of which will be of primary concern here. The nutrients provide for growth and maintenance of the body's tissues and supply it with heat and energy; they also control and regulate many internal processes of the body. Both environmental and individual factors determine the amounts of nutrients needed. Obviously, someone who chops wood all day will need more calories (a calorie is a unit of energy) than someone who sits in an office; an athlete will need more than a librarian.

Carbohydrates are composed of carbon, hydrogen, and oxygen. They provide about 50 percent of the calories in the United States and more than that in many other countries. A gram of carbohydrate yields four calories of energy. The complex carbohydrates include starch and cellulose. Cellulose, a major component of plants, cannot be digested by humans, but is needed for roughage. The simple carbohydrates are the sugars, such as sucrose, or table sugar, which is generally secured from either sugar cane or sugar beet, and glucose, or corn sugar. Lactose, or milk sugar, is the only sugar that comes from animals. Honey is primarily sugar and water; it contains little else of nutritive value, so it is little different from the other sugars in spite of the claims of some health-food faddists. Many nutritionists claim that far too much sugar—over one hundred pounds per person annually—is consumed in the United States. Sugar consumption is not even necessary, for the body converts starches to sugar.

Lipids, the fats and oils, are also composed of carbon, hydrogen, and oxygen, but are a much more concentrated source of energy than are carbohydrates, giving nine calories of energy per gram. The Surgeon General has pointed out that the people of the United States eat far too much fat; some fats or oils, however, are essential in the diet, for they carry the fat-soluble vitamins. The main components of the lipids are fatty acids, of which three classes may be recognized. The saturated fatty acids are the chief constituent of animal fats but are found also in some plants, notably the palms; their consumption raises the level of cholesterol in the blood. The unsaturated fatty acids, found in many plants and in fish, may lower the cholesterol levels, but recent experiments indicate that they may suppress the immune system as well. The monounsaturated fatty acids are also found in plants and may likewise reduce cholesterol levels. Cholesterol is found in both low-density lipoproteins (LDLs) and high-density lipoproteins (HDLs). The former contain the greatest percentage of cholesterol and may be responsible for deposits of cholesterol in the blood vessels. Exercise and other factors such as dietary habits may influence the higher levels of LDLs in some people. The implication of cholesterol in certain diseases, arteriosclerosis, or hardening of the arteries, for example, has led to the increased use of plant oils over those of animal origin. Recently people have turned more to the monounsaturated peanut and olive oils.

Two new fat substitutes have been developed in recent years that can be used in many of the same ways as natural fat. One is low calorie and the other passes through the digestive tract without being absorbed. At this writing, both are still awaiting approval by the U.S. Food and Drug Administration (FDA).

Proteins, which differ from carbohydrates and lipids in that they contain nitrogen in addition to the other elements, are composed of amino acids. Of the twenty or so naturally occurring amino acids, nine—the so-called essential amino acids—must be supplied in the adult diet. Although the specific function of protein is to build and maintain body tissues, it is burned for energy if not enough carbohydrates and lipids are available for that purpose. Protein provides four calories of energy per gram.

Meat is most people's favorite food, and properly so because it provides complete protein, that is, protein that includes all the essential amino acids in the proper proportions for human nutrition. Meat also supplies some vitamins and minerals. Egg is the best source of protein (Table 3–1). although some nutritionists recommend that people eat only two eggs a week because the yolk has a very high cholesterol content. Plants, like all organisms, contain protein, but it is usually incomplete (for humans) in that one or more amino acids are present in insufficient amounts. Nonetheless, the people of the world get most of their protein, as well as their calories, from plants; the cereals are the chief source of both protein and carbohydrates. Meat is much more costly than plant foods, since animals themselves consume large

Table 3–1 Essential amino acid composition (milligrams of amino acid per gram of nitrogen) of certain foods

Food	Iso-leucine	Leucine	Lysine	Methio-nine	Phenylal-anine	Threo-nine	Trypto-phan	Valine	Protein score[a]
Hen's egg	393	551	436	210	358	320	93	428	
Beef	301	507	556	169	275	287	70	313	69
Cow's milk	295	596	487	157	336	278	88	362	60
Chicken	334	460	497	157	250	248	64	318	64
Fish	299	480	569	179	245	286	70	382	70
Corn	230	783	167	120	305	225	44	303	41
Wheat	204	417	179	94	282	183	68	276	44
Rice	238	514	237	145	322	244	78	344	57
Bean	262	476	450	66	326	248	63	287	34
Soybean	284	486	399	79	309	241	80	300	47
Potato	236	377	299	81	251	235	103	292	34
Manioc	175	247	259	83	156	165	72	204	41
Coconut	244	419	220	120	283	212	68	339	55

Source: FAO Nutritional Studies, no. 24, Rome, 1970.

Note: Histidine, which has recently been shown (Laidlaw, S. A., and J. Kopple, *Amer. Jour. Clin. Nutr.* 46 (1987): 593–605) to be an essential amino acid, is not included in the table.

a. Egg is considered to have a nearly ideal protein and the other foods are rated in comparison with egg to give a protein score. Note that protein score is not the same as protein content, which is not included in this table.

amounts of plant materials before it is their turn to appear on the table—it takes roughly seven pounds of grain to make a pound of meat. On the one hand we could feed far more people if meat eating were reduced. On the other, grazing animals such as cattle and sheep are able to utilize food that humans cannot. In the United States and many other developed countries, however, most of the meat-producing animals are fed grain; in fact, about three-fourths of the grain grown in this country is fed to animals. Only in the wealthy countries do people consume large amounts of meat (Table 3–2). Although meat consumption in the United States has fallen from 192 pounds per capita in 1976 to 174 pounds in 1987, this country is still one of the leaders in world meat consumption.

There are, of course, many people who for religious or other reasons do not eat meat. They are sometimes divided into vegetarians, who eat eggs and milk products, and vegans, who do not eat any food of animal origin. The latter will not obtain enough vitamin B_{12} unless they take it as a supplement, for it is found in significant quantities only in foods

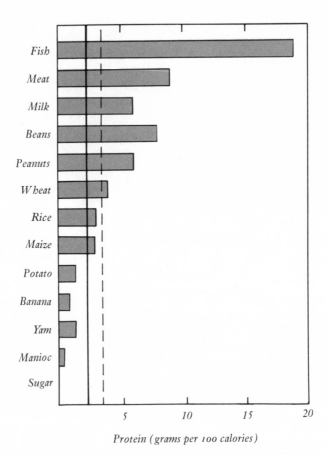

Protein (grams per 100 calories)

Figure 3–1 Protein-calorie ratios of various foods compared. It can be seen that some of the basic foods of tropical areas, such as banana, yam, and manioc, are protein-poor. The approximate adult dietary requirement of protein, in grams per 100 calories of food consumed, is shown by the solid line, that for children by the dashed line.

of animal origin. Adequate vitamin B_{12} is essential for proper growth, production of red blood cells, and functioning of the central nervous system. As for protein, if plant foods are eaten in the proper combinations, a diet supplying the proper proportions of amino acids can be achieved. For example, rice is low in the amino acid lysine but high in the amino acid methionine, while beans are low in methionine but high in lysine. When eaten together, as they are in many parts of the world, rice and beans provide a fairly complete protein; so, for similar reasons, does a peanut butter sandwich. A diet limited to these foods is hardly to be recommended, however.

Table 3–2 Daily per capita consumption of calories and protein from plant and animal sources for selected countries

Country	Calories			Protein (grams)		
	Total	Plant	Animal	Total	Plant	Animal
Bangladesh	1,859	1,795	64	38.6	34.3	4.3
Burundi	2,217	2,166	50	71.3	67.8	3.5
Canada	3,443	2,217	1,226	94.2	36.2	58.1
China	2,564	2,367	198	60.2	51.9	8.4
Ecuador	2,031	1,678	353	45.0	24.0	21.0
France	3,337	2,124	1,214	106.6	36.2	70.4
Ghana	1,679	1,597	81	36.7	26.5	10.2
Haiti	1,843	1,742	101	44.0	37.0	7.0
India	2,161	2,036	125	52.3	46.1	6.3
Japan	2,804	2,263	541	95.6	41.9	43.7
Mexico	3,147	2,604	543	81.3	52.4	28.9
Mozambique	1,664	1,609	56	29.1	25.1	4.0
Peru	2,144	1,874	270	56.8	37.2	19.6
United Kingdom	3,130	2,012	1,118	85.5	34.3	51.2
United States	3,652	2,388	1,264	104.4	35.1	69.3
USSR	3,403	2,536	867	98.3	49.5	48.8

Source: FAO Production Yearbook, vol. 40, Rome, 1986.

Many taboos have grown up concerning the eating of animals, whereas the intake of plant foods seems not to have been regulated in this way. People may think there is a rational origin for their rejection of certain foods, but most food taboos actually arose in prehistoric times and their origins will probably always remain obscure. Most taboos are associated with religion, but in no consistent way. Some religions permit the eating of sacred animals, while others forbid it. It could well be that many taboos did not have their origin in religion, but only later became associated with it. Food taboos are still very much with us. In fact they may at times contribute to protein malnutrition of people who will not eat fish, eggs, or other foods because they are "unclean" or thought to be objectionable in some other way.

CHAPTER FOUR

Meat: The Luxury Food

Let him not eat of either the cow or the ox; for the cow and the ox
doubtless support everything here on earth. The gods spake, "Verily,
the cow and the ox support everything here: come, let us bestow on the
cow and the ox whatever vigour belongs to other species!" Accordingly
they bestowed on the cow and the ox whatever vigour belonged to
other species; and therefore the cow and the ox eat most. Hence, were
one to eat of an ox or a cow, there would be, as it were, an eating of
everything, or, as it were, a going on to the end . . . Nevertheless
Yâgñavalkya said, "I, for one, eat it, provided that it is tender."

SATAPATHA-BRÂHMANA III, 1, 2, 21

The domestication of all of our important animal species occurred quite
early. That other animals which might have been domesticated were not
is probably in part a geographic and historical accident. Most of the
important domesticated animals came from the Near East, a few from
southeast Asia. After they were domesticated, use of these animals
spread around the world. The ancient Egyptians kept a large number of
animals, but with the exception of the cat, none became truly domesticated. A few animals were domesticated in the New World, but with
the exception of the turkey they did not become widely used outside of
their homeland. Columbus brought cattle and sheep with him on his
second voyage and these Old World animals soon became widespread
in the Americas. People for the most part have tried to adapt the same
animals to new habitats rather than to exploit new species, although
recently there have been a few attempts to domesticate other animals
such as the musk ox. The eland, an antelope, is now said to be truly
domesticated in Africa as the result of such an attempt.

A domesticated animal is one that breeds under human control. (If we
accept this definition, it follows that human beings are not a fully domesticated species, for they have not yet succeeded in controlling their
own breeding.) This definition might be quibbled over, but it should
serve. An animal may be tamed, as a wild plant may be cultivated, but
that does not make either of them "domesticated." Some animals whose
feeding and protection have been assumed by humankind have become
so completely domesticated they are no longer able to survive without
human care. But many domesticated animals in suitable environments

later become feral, or wild, as has happened to the horse and pig in the United States.

Present evidence indicates that plant and animal domestication began at approximately the same time in the Near East, although it has frequently been assumed that domestication of plants preceded that of animals. The earliest date known for domesticated sheep antedates those known at present for any cultivated plant. It may be that some hunters and gatherers were domesticating animals while others were concentrating their attention on plants. However some have argued that, even if this were true, plant cultivation and village life would have to have been established before domestication of animals could have proceeded very far. Thus, the great division of farming and pastoral peoples reflected in the Old Testament story of Cain and Abel may have been a late development.

Only about fifty or so animals have been truly domesticated, including the honey bee, the silk moth, and a few aquatic animals such as carp and trout. Of these only a dozen or so are of great importance and have a wide distribution. If we omit dogs and cats, for which there are no reliable estimates, we find that chickens are the most numerous domesticates, with ten billion in the world. Cattle are second and sheep third, each numbering over a billion; these are followed by pigs, with over half a billion representatives, and goats, with slightly under half a billion. The amount of meat produced by the various animals is shown in Figure 4–1.

Today, in many parts of the world, animals are still kept in ways not very different from those of the prehistoric period. There have been radical changes in the United States and some of the other developed nations, however, most of which date back less than half a century. The small barnyard with its variety of animals is disappearing and being replaced by large, specialized farms. There are now automatic feeding

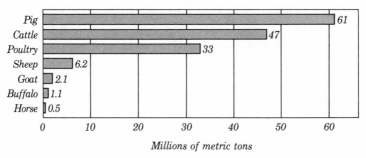

Figure 4–1 Major meat sources for the world in 1986. Poultry includes chickens, ducks, and turkeys. (Figures from *FAO Production Yearbook*, vol. 40, Rome, 1987.)

devices, and the amount and kind of food they dispense are based on the latest nutritional research. Protein concentrates and feeding supplements are used extensively. Modern sanitation practices, immunizations, and treatment of diseases with antibiotics and other drugs have contributed to the production of much healthier animals.* The animals themselves have changed as a result of work done at modern breeding farms, using planned breeding systems based on extensive records and the supervision of trained geneticists. Animal breeding today is based on performance characteristics such as amount and quality of milk, eggs, or meat produced, rather than on appearance as animals are judged at state fairs or animal shows. The new developments, coupled with highly organized processing and marketing systems, have drastically changed the livestock industry in the space of a few years.

Just as they do today, animals served humankind in early times in many ways other than as food. They provided leather and wool for clothing, bones for tools, dung for fertilizer and fuel, and a means of traction and transportation; they served for amusement and religious offerings as well. Today animals are also used in the manufacture of pharmaceuticals, fuel, fertilizers, greases, oil, gelatin, glue, and other industrial products. Catgut, usually made from sheep intestines, is used for violin and tennis strings and for sutures. Chiefly because of the existence of many synthetic products, however, we are far less dependent on animals for raw materials than were primitive peoples. Finally, animals serve us today as subjects in experiments for medical and scientific research.

One group of animals, the even-toed ungulates, or hoofed animals (order Artiodactyla), has provided fifteen of the twenty-two most important domesticated animals. Of these the ruminants, or cud-chewing animals, contributed three of the most important as well as earliest domesticates: cattle, sheep, and goats. These animals, because of the microflora—bacteria and protozoans—present in their stomachs, are able to digest food that humans cannot. In a sense they were preadapted to domestication in that they were not in competition with people, being able to exist on a diet that was utterly worthless to humans. Also, the fact that they were social rather than solitary animals may have made

*But perhaps not always to healthier humans. Some scientists have contended that the increased bacterial immunity to antibiotics used for people stems in part from the indiscriminate use of these antibiotics in animals. The hormone DES (diethylstilbesterol), once widely used to increase growth in steers, has been linked to cancer in humans; its use was banned in the United States in 1979. Other growth-promoting hormones approved by the FDA are now employed. In 1989, however, the European Community prohibited the import of U.S. beef because of the use of these hormones.

their domestication easier. The most important food animals will be treated in greater detail in the following paragraphs. The first to be discussed, although seldom used for food today, may have first been domesticated for that purpose.

Dogs

The dog (*Canis familiaris*)* has generally been regarded as the first domesticate, but only recently has this claim been supported by archaeology. The oldest reported remains of dogs have been found in Iraq and date from twelve to fourteen thousand years ago. Bones dated at 8400 B.C. have been found in western North America. It has generally been assumed that the dog was domesticated in the Old World and came to the Americas with humans in their later migrations across the Bering Strait. It is now fairly clearly established that the wolf was the ancestor of the dog, and some have thought that earliest reported remains of dogs may in fact be of wolves. Wolves had a wide distribution in both the Old World and the New and still have a fairly extensive distribution in spite of human extermination in many places. No one has yet investigated the possibility that the dog in the New World could have been domesticated from indigenous wolves rather than having come from the Old World. Wolves are easily tamed when young, and perhaps people in various places and times brought pups into their camps. The fact that dogs have certain traits in common that distinguish them from wolves does not necessarily mean that they had but a single origin. Dogs have, for example, a curly or sickle-shaped tail in contrast to the drooping tail of their wild relatives. Selection of this characteristic would perhaps have helped people distinguish their own animals from wild ones, and the mutation or mutations that caused it could have happened more than once.

How important dogs were to hunters in early times is the subject of some dispute; some think that they played no great role. Nor is there agreement about whether dogs were important in helping round up and control the herd animals that were the next domesticates. In times when food was in short supply, dogs and humans would have been competi-

*The scientific, or species, name of a plant or animal is composed of two words: the genus name (*Canis*, in our example) and a modifying epithet (in this case, *familiaris*). Thus, *Canis familiaris* is the name of a species, or specific kind of organism, that is commonly called the dog. A genus (plural, genera) is composed of species that are closely related and have several characteristics in common. For example, the gray wolf (*Canis lupus*) is another species belonging to the genus *Canis*. Species may be subdivided into races, varieties, or subspecies. Genera are grouped into families, families into orders, and so on.

Figure 4–2
Clay dog, Colima, Mexico. Dogs for eating were fattened on maize. (Original in National Museum of Anthropology, Mexico).

tors. The dog is a scavenger, however, and in time this trait became appreciated; it is still the dog that helps to keep the village clean in many areas of the world. In parts of both the Near East and South America today, dogs feed largely on human feces. The dog became prized as a source of food in many cultures. In Mexico dogs of a special breed grown for eating were castrated for fattening. In parts of China the dog is still used for food. During the Mao regime there was a conscious attempt to eliminate dogs because they competed for food with people, so as a pet the dog largely disappeared from China. Apparently there are still plenty of dogs in China, however, as is attested by the following article which appeared in the *Times* (London) on January 3, 1980.

> Peking, Jan. 3—A restaurant in Jilin, north-east China, was praised by the *People's Daily* for capitalist-style enterprise in ensuring supplies of its most popular item—dog meat.
> It appealed to people to bring in their own dogs to be eaten and it would buy them. The result: in under a month, it bought 1,369 dogs—a year's supply.—UPI.

The dog became the most widely distributed domesticated animal because it was appreciated and respected as an aid in hunting, in keeping the flocks, and as a scavenger, as well as being used occasionally as a draft animal or as a source of fur or food. Although in some parts of the world the dog still earns its keep in these various ways, its role is becoming more and more that of favorite pet. As such, dogs' food consumption and their contribution to pollution in large cities such as New York have been newsworthy in recent years. It has been estimated that nearly $2 billion a year is spent on canned or packaged dog and cat food in the United States. It has also been pointed out that pets in the United States are better fed than people in some parts of the world. The Chinese perhaps had the right idea; the elimination of pets might contribute toward the eradication of hunger. Dog owners are unlikely to find this acceptable, and they might justify their position by claiming that the need for psychiatric help would increase greatly if pets were prohibited.

Sheep

The ruminants of the family Bovideae include sheep, goats, cattle, and the water buffalo—animals valued highly not only for meat but also for their milk and for their skins or wool. Sheep (*Ovis aries*) are the first of these animals to appear as domesticates in the archaeological record, being represented by remains from 9000 B.C.

Several kinds of wild sheep known as urials or mouflons occur in Asia and Europe, and it is thought that the Asiatic mouflon (*Ovis orientalis*) is the most likely ancestor of the domesticate. The bighorn sheep (*Ovis canadensis*) of North America was never domesticated. Today most of the wild sheep of the Old World are mountain animals, although they probably existed in lower regions in earlier times. Most of the bones uncovered come from the Near East, and it is not always clear whether the bones found in the earliest archaeological deposits are from domes-

Figure 4–3   Sheep in Bolivian highlands. (Courtesy of FAO.)

Figure 4–4 Fat-tailed sheep with cart for carrying tail, from Rudolph the Elder's *New History of Ethiopia*, A.D. 1682. (From Frederick E. Zeuner, *A History of Domesticated Animals*. London: Hutchinson, 1963. Used by permission.)

ticated or wild animals. With domestication came changes in the ear, the face, the horns (which in some races disappeared entirely), the color and other characteristics of the wool, and the tail—mostly traits of little help to the archaeologist in distinguishing the bones of wild and domestic forms. A large tail with an abundance of fat was much appreciated by early people, and apparently there was deliberate selection of this feature. Some sheep developed tails so heavy that carts were constructed to help them carry these appendages around. Fat-rumped sheep were also selected in early times, as animal oils were highly valued for use in lamps. Wool may have been made into felt cloth long before it was spun and woven; perhaps the spinning and weaving of wool were first done in places where plant fibers such as flax were not available. Sheep are unique among domestic animals in their adaptation to environments that are extremely unfavorable for other animal species. They probably became more widespread than goats in early times because they could survive better in hot climates. They have a panting mechanism that allows them to tolerate heat. Surprisingly their wool may also function as a cooling device in sunny desert areas, although most authorities consider it only an adaptation for cold.

Today, as is true for nearly all of our domesticated animals, there are numerous breeds of sheep; some are specialized as producers of wool, others for meat, and still others for milk. Most breeds of sheep in the United States, however, are used for both wool and meat. Cattle ranchers are often opposed to sheep production because the grazing habits of sheep can destroy the rangeland and their fine hooves tend to ruin the watering places. Despite the fact that few products are the equal of wool for making clothing, the use of wool has declined in recent years as synthetic fibers have become widely available. Synthetic fibers are manufactured from petroleum, however, and as the price of petroleum increases, wool may again be in great demand.

Goats

Although dated remains of the first known domesticated sheep are fifteen hundred years older than those of goats, this does not necessarily indicate that the goat (*Capra hircus*) was domesticated later. In many deposits bones are present that could have belonged to either goats or sheep; positive identification is not always a simple matter. In many places where remains of both animals are found in the same deposits, the goat remains appear older than those of the sheep. Wild goats, or bezoars, the now quite rare ancestral form, once extended across southern Asia from India to Crete. The original domestication could have been in the Near East, both Iran and Iraq having been suggested. The goat, in contrast to the other domesticated animals, is a browser, even at times climbing trees to get at leaves, and it can survive in areas where the food supply is inadequate for other animals. It also is able to graze

Figure 4–5   An Angora buck. (Courtesy of *Sheep and Goat Raiser.*)

Figure 4–6   Goats at watering place in Mauritania. (Courtesy of FAO.)

and may eat grass so close to the roots as to promote erosion in seasonally dry areas. There is some controversy, however, about how detrimental the goat is to the land. Although goats have been generally condemned for their destructive grazing habits and have converted some grasslands to deserts, they also have their defenders, who point out that goats are often put to pasture where cattle have already grazed and thus at times may be blamed for erosion actually started by the cattle. It has been pointed out, moreover, that shrubs may invade and ruin grasslands overgrazed by cattle, whereas goats protect such grasslands by eating shrubs.

Domesticated goats spread rapidly but were nearly always less appreciated than either sheep, which are far superior for wool production, or cattle, which are superior for milk production. In addition, both sheep and cattle were generally preferred over goats for meat. Goat production is not very significant today, except in certain steppe and mountainous areas and in some parts of Africa and Asia where goats are still more important than sheep. At present nearly half of the world's goats are found in Africa. One breed, the Angora goat, is still important for its wool, which is called mohair. Goat's milk, which in some ways is more similar to human milk than is the cow's, still has some use. The fat globules are small in size and more easily digested than those of cow's milk. It is rather difficult, however, to imagine people who had either sheep or cattle domesticating goats, which might support the argument for goats being the first ruminants to be domesticated, or for their being domesticated in areas somewhat removed from centers of sheep and cattle domestication.

Cattle

Generally considered the aristocrats of the domesticated animals, cattle (*Bos taurus*), so far as we know today, were domesticated later than sheep and goats. Cattle first appear in the archaeological record shortly after the appearance of goats, but they do not become common until much later. Whether success with other animals inspired attempts at domestication of cattle or whether they were domesticated by people unacquainted with other domesticated animals is not known. The aurochs, or wild cattle, were worshiped long before their domestication, and a religious motive for their domestication has been postulated. These animals had, of course, been hunted in earlier times, as some magnificent cave drawings in Europe attest. The aurochs, the last of which were killed in Poland around A.D. 1630, were magnificent animals. Cattle thought to be similar to some of these original wild ones have been developed in modern times in Germany by the interbreeding of modern types with certain presumably primitive characteristics. The animals ultimately produced were large, strong, temperamental, and ferocious, but quite agile, unlike the thickset beasts most common today. The domestication of the aurochs was certainly not a simple matter. Salt may have been used to entice the animals near human habitations to allow the capture of young animals. One of the changes frequently accompanying domestication in animals is a decrease in size from that of the progenitor. In cattle this may well have resulted from an intentional selection of smaller beasts that could be more easily managed. It could also simply be due to an environmental factor, in that the animals may not have eaten as well under domestication as they had in their natural environment.

The aurochs were widely distributed throughout the temperate parts of Europe, Asia, and North Africa. The earliest known remains of clearly domesticated cattle, dated at 6400 B.C., come from Turkey. Although

Figure 4–7 The aurochs, based on a drawing of the last surviving specimen. (From Frederick E. Zeuner, *A History of Domesticated Animals*. London: Hutchinson, 1963. Used by permission.)

some have considered that humped cattle (*Bos indicus*), sometimes treated as a separate species, originated in India, present authorities maintain that it is likely that the hump developed through artificial selection of ordinary cattle in Iraq and went from there to India. Crosses between the two types of cattle are readily made. Various breeds developed quite early. As cattle were introduced into new areas there was probably mating of the cows with wild bulls, either naturally or by human design. Primitive people are thought to have staked out female animals, a practice still followed with reindeer, to entice male animals, which are then captured or killed for food. The introduction of genes from the mating of wild bulls with domesticated stock probably contributed to the early development of considerable diversity in cattle and was the foundation of new breeds.

We do not know when it was first discovered that castration could have a profound effect on the bull,* rendering it docile and manageable.

*People also quite early practiced castration on other animals, including themselves. The most widely practiced method, now as then, is to open the scrotum with a knife and remove the testicles, but some people castrated their animals by pounding the testicles between stones. In addition to changing the metabolism and behavior of the male animals, castration obviously was a method of birth control. There were probably good reasons at times to restrict the breeding of animals, such as the fact that young born at certain times of the year in adverse environments might have little chance of survival. Other methods of birth control included fitting leather aprons to the animals or binding their prepuces with string. Both of these methods are still used with rams in Africa.

A religious origin as a sacrifice of the male element, among people who worshiped cattle, has been suggested, but it is perhaps more likely that it was a purely practical matter. Cattle raisers would have found that keeping more than one bull in a herd created difficulties that could be solved by castrating all but one of them. The ox, a castrated male used for work, was one of the first beasts of burden and is still very important in parts of the Old World and Latin America. It made possible the plowing of fields after the plow came into use, perhaps around 3000 B.C., and thus had a profound effect on the development of agriculture by considerably extending the area that could be tilled.

People must have used milk for food soon after the herd animals were domesticated. Although milk is regarded in some parts of the world as one of the best foods, there are many adults who cannot tolerate it, as was learned when the United States sent powdered milk as part of relief shipments to various countries. For lactose, or milk sugar, to be digested, it must be broken down into simpler sugars by an enzyme called lactase. Virtually all babies produce lactase, which enables them to digest their mothers' milk. In the past it was thought that humans lose the ability to digest lactose if they are not continually fed milk after weaning, but it has recently been pointed out that there may be a genetic difference among people for retaining this ability. It is largely adults of the nondairying cultures of Africa and eastern Asia and of Native American groups who are unable to digest lactose. These people can, however, eat fermented milk products such as yogurt and cheese, which have a lower lactose content.

In parts of Africa, where they became a symbol of wealth, cattle were used for currency, and dowries are still frequently made up of cattle. In some areas of Africa cattle are seldom or never eaten, although their blood and milk may be used for human food. Members of the Masai tribe obtain blood by shooting an arrow at close range into the vein of the neck of an animal, then collecting the blood in a gourd. The Abahima of Uganda have selected their cattle for large horns, which may reach a weight of 150 pounds, and a hump so large that it droops over to one side. The appearance of these animals is pleasing to the Abahima, but the large horns and humps represent no economic benefit and in fact may be detrimental to the well-being of the cattle.

The role of cattle in India has received considerable publicity and is still a highly emotional issue there. Cattle are sacred among the Hindus, and it has been said that they would rather die of starvation than kill their animals. This failure to use cattle as a source of meat has been much criticized by outsiders. The origin of the Hindu taboo on eating cattle is obscure. That the animal was sacred may or may not have much

Figure 4–8 Ancholi cattle of Burundi. The long horns confer no economic advantages. (Courtesy of FAO.)

to do with it, for some peoples do eat the animals they hold sacred. It has been suggested that the animals were so valuable as a source of traction and in providing milk that there was an early prohibition on killing them. Later, of course, the taboo may have been reinforced by the entry of cattle-eating foreigners, first the Moslems and later the British. India is overpopulated with cattle; nearly one-sixth of the world's cattle are found there, many freely roaming the streets. The animals do make many contributions: the cows provide milk, the bullocks are the principal source of traction, and the dung is used as fuel for cooking and as construction plaster. It has been pointed out, however, that in India fifteen cows are needed to produce as much milk as does one in the United States and that the Indians cannot get efficient labor from their ill-fed bullocks. Many people feel that the Indians would be far better off with fewer, more productive animals. Marvin Harris says instead that cattle are no more sacred in India than is the automobile in the United States and claims that there is a sound ecological basis for their status in India. Indian cattle are scavengers and do not compete with humans for food, and in terms of energy India makes better use of cattle than does the United States. Certain lower-class Hindus, the "untouchables," do eat the meat and also take the hides. Cattle are also sold to Moslem traders who in turn sell them for meat. There are few cattle in India that do not end up in the pot.

The Indian cattle, or zebu, sometimes known as Brahman cattle in the United States, differ from other cattle in several respects. In addition to having humps, they have long heads, drooping ears, loose skin, and longer legs; they are well adapted to hot climatic zones that have a pronounced dry season. Zebus have been extensively used in this century for breeding purposes in the southern United States and in Latin America to provide more productive cattle for these areas.

The value attached to cattle as food has led to their introduction into many parts of the world where they are not well adapted, the semiarid scrub region of the western United States, for example. Cattle cannot be grown in some parts of Africa because of parasites, chiefly tsetse flies. Various diseases still plague cattle in other parts of the world, hoof-and-mouth disease being one of the most serious. An outbreak of this occurred in England in 1967 and more than four hundred thousand cattle had to be destroyed and burned, representing a loss of $250 million.

Most cattle in the United States today are bred either for milk or beef, with only a small percentage having a dual purpose. There have been

Figure 4–9 Zebu bull in Burma. These Asian cattle, which have been found superior to other cattle in many tropical and subtropical areas, have been widely used for breeding purposes. (Courtesy of FAO).

great advances in dairy science in the last half century, and milk production per cow has greatly increased. Milking machines are now widely utilized, and the freezing of semen and artificial insemination are extensively employed in breeding. Transplanting embryos from superior cows to average ones has proved successful in producing larger numbers of highly productive cows. High butterfat content, once considered very important, has been de-emphasized because of consumer concern over calorie and cholesterol intake; moreover, numerous plant oils are available for the manufacture of butter substitutes. Although there has been no shortage of milk in the United States, a bovine growth hormone, produced with the aid of biotechnology, is available that will greatly increase milk yield. Its use has not yet been approved by the FDA, however. Modern transportation and improved breeding and feeding practices have also made cattle far more efficient meat producers than in the past, but beef still ranks as the most expensive meat in terms of the cost of feeding the animals. In has been calculated that 80 percent of the grain produced in the United States is fed to animals, much of which goes to fattening cattle. Even though leaner cattle are now being produced in greater numbers than before, the National Academy of Sciences recommends doing more. Feeding cattle more on grass and less on grain leads to a reduction in the content of fat. At the same time, less fat makes the meat less tender and flavorful, two of the very reasons beef has become our most popular meat. In part because of worry over fat and cholesterol, beef consumption declined by 12 percent in the last decade. Consequently, the U.S. Beef Industry Council has staged an ambitious advertising program—$25 to $30 million dollars being spent in 1988—to promote the eating of beef.

Beefalo

In recent years the beefalo has received considerable promotion in the United States. This animal, originally known as cattalo, is a hybrid between cattle and the American buffalo, or bison (*Bison bison*), another member of the bovid family; it has been known since the latter part of the last century. It was found that the bison had to serve as the female parent, for when the domestic cow was used, the broad shoulders of the calf, inherited from the bison, made birth difficult, often killing the mother. Some problems of sterility in the hybrids were encountered, but most of these seem to have been overcome. The animals raised for the market result from backcrossing the hybrids to cows. Claims have been made that beefalo are superior to cattle in a number of respects, including faster growth, a higher percentage of protein, less fat, a greater ability to withstand temperature extremes, and resistance to many diseases that affect cattle; some have disputed these claims, however. So far beefalo have made little impact on meat supplies in the United States, but of course they may contribute more substantially in the future. These animals, however, are unlikely to be the solution to feeding a hungry world as some of their proponents have claimed.

Buffalo

Also a member of the bovid family, the Indian, or water, buffalo (*Bubalus bubalus*) is still extremely important in many parts of the world. The other buffalo and bison of Europe, North America, and Africa were never domesticated. We know little more about the domestication of the

Figure 4–10 A dairy herd of buffalo near Calcutta. The buffalo must spend considerable time in water. (Courtesy of FAO.)

Figure 4–11 Buffalo plowing a paddy in Indonesia. (Courtesy of FAO).

water buffalo than that it occurred in southeastern Asia, presumably sometime before 2500 B.C. A few of the wild species, the arni, or Indian, wild buffalo that gave rise to the domesticate, still exist. Two main types of the domesticate, the swamp and river buffalo, are recognized. Both thrive in tropical lowlands, like to wallow in water, and feed largely on aquatic or semiaquatic grasses and other vegetation. Their use is similar to that of cattle, and in many cultures they occupy the same ceremonial role as cattle. They are a good producer of milk, which has 7 to 10 percent butterfat content, higher than that of most cattle. About 70 percent of the milk used in India and Pakistan today comes from buffalo. Butter made from buffalo milk is greenish-white in color. It is also more solid than cow's butter and turns rancid less readily. Buffalo meat has a flavor very similar to that of beef, but it has a distinct bluish tinge and fat that is white. Thus far there has been little scientific breeding to improve the animal.

The water buffalo spread throughout most of southeastern Asia, where it became important in preparing fields for growing rice. It is in fact the only animal well adapted to work in muddy fields. The late arrival of the animal in Africa and Europe is rather puzzling; it didn't reach Italy, where today its milk is used in making mozzarella cheese, until about A.D. 700. It is also raised today in other parts of southeastern Europe, Egypt, and the Middle East. The water buffalo was brought to Brazil in 1903, where it has become important in the lower Amazon valley. Efforts to introduce it in many parts of Africa have failed. The animals have escaped and become feral in northern Australia, as also has happened in South America and in their homeland. There are sometimes reports of hybrids between water buffalo and cattle, but these have not been scientifically verified.

Horses

Most readers of this book will never have eaten horse meat, at least not knowingly. Occasional "scandals" have stirred in the past when it has been found that horse meat was substituted for beef. There is nothing wrong with horse meat: wild horses were common game for our Paleolithic ancestors, and horse meat is still eaten today by many people in central Asia, the area where the horse (*Equus caballus*) was domesticated. The milk of the horse is employed for making kumiss and other fermented beverages. Why, then, do so many people refuse to eat horse flesh? As with all food taboos, the origin of this avoidance is obscure, and there probably is no logical reason. Various suggestions have been put forward. The horse, more than any other animal except dogs and cats, was friend and close companion to humans, and people don't usually eat their close friends. Perhaps the animal was so useful in other ways—in agriculture, transportation, or warfare—that its eating was discouraged. Possibly Christianity played a role in the rejection of horse meat in Europe, since the eating of it was associated with pagans. There were attempts to popularize horse meat in Europe in the last century, but they largely failed except in France. In 1969 quite an outcry was raised in England when the people learned that retired horses from the queen's pound were being sold in France for use as food. As a result the government changed its policy, promising to put retired horses out to pasture. It seems unlikely that more widespread use of horse meat would do much to ease hunger in the world, since horses are even more expensive eaters than cattle. A few years ago horse meat, much of it from feral horses, was used in the United States for feeding dogs and cats, but it is now too expensive for that purpose. What meat is now canned in the United States is shipped abroad.

The earliest archaeological evidence of the domesticated horse is from the southern Ukraine and dated at 3500 B.C. It is generally considered that the tarpan, a wild horse that became extinct in the last century, was its ancestor. Recently it has been postulated that a second domestication

Figure 4–12 Donkeys used to haul grain. From an Egyptian tomb, c. 2400 B.C. (From Charles Singer et al., eds., *History of Technology*, vol. I. London: Oxford University Press, 1954.)

of the horse occurred in northwestern China around 2000 B.C. Its ancestor appears to have been the wild Mongolian race, or the Przewalski horse, now extinct in the wild but still found in zoological parks.

Though the domestication of the horse brought many benefits, it had some unfortunate consequences as well. Once they were provided with horses, nomadic peoples became a great scourge to the sedentary farmers. From 2000 B.C. on, mounted warriors and horse-drawn chariots "swept across the western world," as Zeuner expresses it. The horse made it possible for the Huns to build their great empire, and it promoted the many martial successes of the Arabs. It was later responsible for the Mongol conquests under Genghis Khan, and it helped the Spanish to conquer the Americas with great rapidity. Some Native Americans adopted the horse soon after its introduction by the Spanish, and it became important to them in hunting and warfare. Until this century the horse continued to occupy a major role in war. Two other late domesticates, the camel and the elephant, were also used in warfare, but neither was ever as important as the horse.

Another species of *Equus*, the ass, or donkey, was also domesticated but it never became as closely associated with people as did the horse. The mule, which results from the mating of a mare and an ass,* was to become the first documented interspecific (between species) hybrid on record. The mule has been described as "an animal without pride of ancestry nor hope of descendants," which isn't quite true, for there are a few records of mules producing offspring. Throughout history the mule has been cursed for its disposition but praised for its sure footedness, and it has made its contributions to agriculture.

*The reciprocal hybrid, from the mating of an ass with a stallion, is called a hinny.

In 1950 it was estimated that 86 percent of the draft power of the world's agriculture, and 25 percent for the United States, was provided by animals. The estimate for the United States seems too high, and of course the use of animals for draft power has drastically declined in this country since then, although for the world as a whole it is decreasing much more slowly. In 1918 there were twenty seven million horses in the United States. This figure declined to three million in 1960 but has nearly tripled since. Although their use in agriculture has drastically decreased, their use in recreation and sports is on the increase. Horses are not always cheap. In 1983 there was a report of a racehorse colt selling for over $10 million, and some stallions are said to be worth more than $100 million. Stud rights for a single breeding have gone as high as $800,000.

Pigs

Pigs (*Sus scrofa*), like dogs, are scavengers and hence were in more direct competition with humans for food than were the grazing animals. Also, unlike many of our other domesticated animals, the pig does not have an important dual role, being kept almost solely as a meat producer. The prolific pig was, and still is, a wonderful supplier of meat—the most productive of food per unit area of the larger domesticated animals. Although at one time more pork was consumed in the United States than beef, that is no longer true. In fact, in recent years chicken consumption has increased at the expense of pork. It has been said that in the modern meat industry every part of the pig is used except the squeal. Its hide, of course, has long been appreciated, "pig-skin" being synonymous with football in the United States. The scavenging habit of the pig may have been important among primitive people in preparing the land for farming. Pigs probably even helped open the forests of Europe for crop planting. The ancient Egyptians employed pigs or sheep to tread seed into the ground during planting. At times pigs have been used for traction. One of their most interesting specialized uses is in France, where pigs, like dogs, have been trained to hunt for truffles. It has also been reported that pigs have been used for retrieving game.

Wild pigs are native to Europe, North Africa, and Asia. The wild pigs of the United States are escapes from domestication that have reverted to a feral existence. Again it appears that the earliest domestication was in the Near East, for pig remains dated at 7000 B.C. are known from Turkey. Some have postulated a separate domestication in China. Pigs were probably not difficult animals to domesticate. The domestication of the pig was somewhat later than that of sheep and goats. This may be due to the fact that pigs could not have been domesticated until there were well-established villages, since the pig is primarily a household animal, not a herd animal. The fact that it could not readily be herded may have led to the development of taboos against the pig among nomadic peoples such as the Jews and Moslems. To them it could have become a despised animal because it was available only to settled people, who were at the same time people with alien gods. This explanation for the origin of the taboo is not proved, of course, but it seems a more reasonable one than others that have been proposed. It seems very unlikely that primitive people could have known that pork could carry trichinosis, or that they would have rejected the animal because it was a scavenger. Mohammed influenced the rejection of pork by his followers, perhaps to make a distinction between them and the Christians, who were pork eaters. The pig was the most important sacrificial animal in ancient Rome, and it is still important in religious observations in parts of southeastern Asia and the Pacific, where the tusks often have special ceremonial significance. Among some tribes of New Guinea pigs are eaten only on ceremonial occasions.

Just as was the case with cattle in Africa and Asia, religious beliefs may impede the improvement of pigs in some parts of the world. In an attempt to improve the food supply of the people in the area of Upper Burma, agricultural agents hybridized the local black pig with a higher-yielding spotted strain. Because the hybrids were spotted, the experiment was a failure, for the local people believed that spotted pigs were unfit for human consumption.

The modern thickset pig most widely raised in North America is derived from crosses of European and Chinese breeds first made in England more than 150 years ago. The smelly pigsty is disappearing today as large modern farms take over pork and bacon production in the United States. The widespread idea that the pig is a dirty animal is

Figure 4–13 Sheep used to tread in seed. The sower (right) offers grain to the animals to keep them following him. From an Egyptian tomb, c. 2400 B.C. (From Charles Singer et al., eds., *History of Technology*, vol. I. London: Oxford University Press, 1954.)

Figure 4–14 An authority on animal husbandry from the Food and Agricultural Organization of the United Nations inspecting a boar in Burundi. (Courtesy of FAO.)

justified only for warm regions, where it wallows in mud or urine in an attempt to keep cool. In cool climates, however, the pig is one of the cleanest of animals. Corn was found to be an ideal fattener for pigs, and therefore it is no surprise that the corn belt of the north central United States is the chief area of swine production in this country. Castration, a very simple operation with pigs, is widely practiced for purposes of fattening.* Boars intended for meat are castrated a few months before slaughter in order to eliminate an undesirable odor from their meat. The breeding of pigs in recent years has produced many changes to meet consumer requirements. People now want less fat than before, plus lard is no longer in great demand.

Chickens

Of the domestic fowl, chickens (*Gallus gallus*) are more important than all of the others combined, and they are one of the few important domesticated animals to have come from the Far East. The chicken is

---

*The Tsembaga of New Guinea castrate all of their male pigs, which means that they have to depend upon feral males to perpetuate the race.

A

B

Figure 4–15 *A.* Chickens, along with pigeons, turkeys and goats, being fed in a farmyard in Togo. (Courtesy of FAO.) *B.* Collecting eggs in a modern poultry farm in Japan. (Courtesy of FAO.)

derived from the jungle fowl of India. It has been postulated that the fowl was originally domesticated to increase its availability for use in divination, rather than for food. Attempting to foretell the future by examining chicken entrails or the perforations of the thigh bone is still practiced in some parts of southeastern Asia. This is also the area where cock fighting apparently originated. Cock fighting was enjoyed in ancient Greece and is popular today in various parts of the world.

Although India is usually regarded as the place of domestication of the chicken, there is as yet no supporting archaeological evidence. After 2000 b.c. chickens reached Iran, Egypt, and China; they became known in Europe more than a thousand years later. The chicken has generally been considered a post-Columbian introduction to the Americas, but the geographer George F. Carter maintains, on the basis of early literature and linguistic evidence, that it was fairly widespread there when the Spanish arrived; he has concluded that it reached the Americas from across the Pacific. Archaeological evidence confirming the early presence of chickens in the Americas has yet to be found, however. In Europe the cock was valued (and perhaps also reviled) early on as a time clock, and its crowing was thought to frighten away the evil spirits of the night. In some areas the chicken became an erotic and fertility symbol, the former because of the cock's elaborate courtship behavior, the latter because of the hen's abundant egg laying. At the same time some people in Africa avoided the chicken because they believed that eating it or its eggs would destroy their sexual functioning and fertility.

The chicken continues to serve as a dual-purpose animal in many parts of the world; in many places the birds are scrawny, largely scavenging for an existence, and producing but few eggs, which are rarely eaten by the owners but taken to the market for sale. More and more, the production of eggs and of "broilers" are separate operations in the industrialized nations, where the raising of chickens depends on procedures in striking contrast to the old barnyard methods. In fact, the broiler farm may be compared to a modern factory with its assembly lines. Mammoth incubators, artificial lighting, and automatic feeding, watering, ventilating and cleaning, and defeathering machinery are utilized in the production of fowl for the supermarket or the chicken-house chain restaurant. The breeding of poultry is more highly developed than that of any other group of animals, and hybrid chickens have come to be of great importance in recent years.

Other Old World Domesticates

There are, of course, several other animals—camels, elephants, and reindeer, for example—that are used in the Old World. None of them, however, figured as prominently in early agriculture as did those already treated. Nor are these other animals as important for food sources today as some of those previously discussed. This is not to say, however, that some of the other domesticated animals—the rabbit, for example, which can be raised very economically—cannot make significant contributions in the future by supplying much needed protein in various parts of the world.

New World Domesticates

In the Americas only a very few species of animals were domesticated, and none of these, with the exception of the turkey (*Meleagris gallopavo*), has become significant outside of the New World. Wild turkeys, now rare, were once fairly widely distributed in the United States and Mexico, and the turkey was already well domesticated in Mexico when the Spanish arrived. There may have been a separate domestication of the turkey among the Pueblo in the southwestern United States. Why the name turkey was adopted by the English is unknown. It may be that the American bird was confused with the turkeycock, or peafowl, or because at the time of its introduction many foreign things were coming from Turkey. Changes in the turkey to produce the broad-breasted bird now favored in commerce has led to the inability of turkeys to copulate naturally; hence, artificial insemination must be employed.

In South America four animals were domesticated in prehistoric times. The Muscovy duck (*Carina moschata*)—why it is called Muscovy has never been entirely satisfactorily explained—is the other fowl that came from the Americas. The guinea pig, cavy, or "cui" (*Cavia porcellus*), a rodent now widely used as a laboratory experimental animal, was domesticated in the Andes and is still an important food animal among the Indians there. It lives in their houses with them and often appears for sale in the market either alive or roasted. The name is thought to stem from its arrival in Europe from Guinea, the animal having gone first from South America to West Africa. More important than either of these species, however, are the llama (*Lama glama*) and the alpaca (*Lama pacos*), both of the camel family, and probably derived from the wild guanaco. A relative of these, the vicuña, much prized for its wool, was never domesticated, although attempts are now being made to domesticate it in Peru.

Llama remains dated about 3000 B.C. have been reported at a site of early agriculture in Peru; the original domestication was probably in highland Peru. Llamas played many important roles among the Andean Indians. They were used for sacrifice, pure black or white animals being

Figure 4–16   Roasted guinea pigs for sale in a market in Quito, Ecuador.

preferred for this purpose. Their lungs and entrails were examined for omens, and potatoes were ritually treated with their blood before planting. Consequently, a religious motive for their domestication has been suggested. The animal was used as a beast of burden, for food, as a source of wool, hide, and medicine, and its dung was used for fuel, but it was neither ridden nor milked. As a beast of burden the llama can carry only a relatively small load, a little more than a hundred pounds, and travel ten to eighteen miles a day. A great advantage is that it can live off the sparse fodder provided by the Andean highlands. Several thousand llamas are now found in the United States, where in recent years they have become popular as pets or pack animals. The alpaca, which is slightly smaller than the llama, is used for most of the same purposes, but not as a beast of burden. Its wool, however, is far superior to that of the llama.

Figure 4–17 Llamas in Bolivia. (Courtesy of FAO.).

After the arrival of the Spanish, the Old World domesticated animals spread throughout the Americas. In Peru it is sometimes said that it took three other animals to replace the llama—cattle for food, the donkey as beast of burden, and sheep for wool—and that all of them required more food than the llama. Both the llama and the alpaca have persisted in the high Andes, where they are well adapted, and will probably continue to thrive there for some time. Although their range is considerably more restricted than it was at the height of the Inca empire, they are a common sight in parts of highland Peru and Bolivia, where they are still used much as they were in the past and are also gaily decorated for ceremonial occasions.

CHAPTER FIVE

Grasses: The Staff of Life

All flesh is grass.

ISAIAH 40:6

Of all the various plant groups—algae, mosses, ferns, gymnosperms,* and flowering plants—flowering plants have furnished us with nearly all of the species we use for food and clothing and in countless other ways. Since the flowering plants, or angiosperms, comprise nearly two-thirds of all species of plants and are the dominant vegetation on the earth's land surface, their great use perhaps should not come as a surprise; but their significance does not derive as much from their numbers as from the fact that they are the only plants that produce fruits and seeds. Of the two hundred thousand or more species of known flowering plants, only three thousand or so have been used to any extent by humans for food. Of these about two hundred have become more or less domesticated, of which some dozen or so are the primary foods that stand between us and starvation. The grasses are foremost in this regard (Figure 5–1).† Of the three hundred or so families of flowering plants, none is of greater importance to us than the grass family, known scientifically as the family Gramineae.

From the time of the earliest seed collectors to the present, the grasses have supplied us with one of our principal foods. Their fruits, or grains, often called seeds, each develop from the ovary of a single flower and each contains a single true seed. Inside the seed coat is a rich layer of stored food, mostly starch, known as the endosperm. This concentrated reserve food, which is designed to provide energy for the germinating embryo, is also a rich source of carbohydrates for us and other animals. The embryo, sometimes called the germ of the seed, contains protein

---

*The gymnosperms, which include the conifers, such as pine, are particularly important to us for their wood.

†Quite a number of plants are sometimes called grass, including marijuana, that do not belong to the botanical family of this name. The true grasses are distinguished from other plants by a combination of floral and vegetative features. Their small flowers, which lack petals, are enclosed in specialized scales, or bracts. The flowers are grouped in small clusters, or spikelets, and the spikelets, in turn, are arranged in clusters, for example, a head of wheat or a corn tassel.

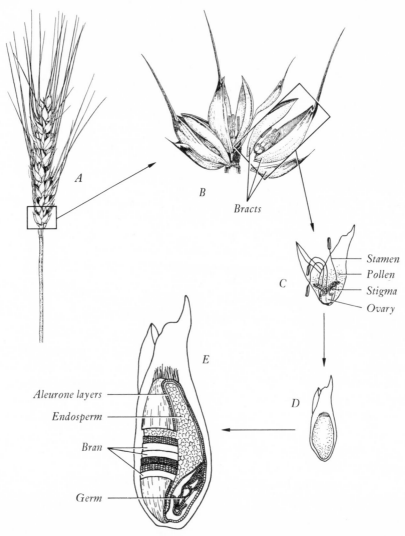

Figure 5–1 Flowers and grain of wheat. *A.* Head, or spike, composed of many spikelets. *B.* Spikelet with three flowers. *C.* Single flower showing parts. *D.* Maturing ovary, or young grain. *E.* Mature fruit, or grain, surrounded by bracts, or chaff; dissected to show various parts. The outer layers of the grain, the bran, contain some carbohydrates, B vitamins, and minerals. The outer part of the endosperm, called the aleurone layer, contains protein and phosphorus. The endosperm, or food layer, is composed mostly of carbohydrates; this is the only part used in making highly refined flour. The germ, or embryo, is rich in fats, protein, and vitamins and contains some minerals and carbohydrates. (*A-D* from "Hybrid Wheat" by Byrd C. Curtis and David R. Johnston. Copyright © 1969 by Scientific American, Inc. All rights reserved. *E* from Wheat Flour Institute.)

Table 5–1 Estimated production of world's major food crops

Crop	Million metric tons	Crop	Million metric tons
Sugar cane*	932	Apple	40
Wheat*	536	Coconut	39
Corn*	481	Cabbage	38
Rice*	476	Rye*	32
Potato	309	Millet*	31
Sugar beet	286	Watermelon	28
Barley*	180	Yam	27
Manioc	137	Onion	25
Sweet potato	110	Sunflower	21
Soybean	95	Rape	20
Sorghum*	71	Bean (dry)	15
Banana and plantain	68	Pea (dry)	14
Grape	67	Mango	14
Tomato	60	Avocado	10
Oat*	48	Pineapple	10
Orange	41	Olive	9

Source: FAO Production Yearbook, vol. 40, Rome, 1986.

Note: The importance of members of the grass family (indicated by an asterisk) is obvious. Considerable amounts of some of the crops listed are not used as food. Most of the yield of grapes goes into the production of wine. The water content of the material should be taken into account in evaluating the overall importance of the various crops. Sugar cane contains far more water than the cereals, and the fleshy fruits such as tomato and watermelon are more than 90 percent water. The millets represent species belonging to several different genera and thus are not directly comparable with the other entries, which represent either a single species or several species belonging to a single genus.

and oil. Some vitamins and minerals are also present in the grain. Thus, the grasses come close to being the ideal source of plant food. Alone they can not sustain us very well, however, for their protein does not contain all the amino acids in the necessary proportions essential for human well-being. The cereals are also deficient in calcium and, except for the yellow form of maize, in vitamin A, and the dried seeds do not contain vitamin C. The grains give high yields, are fairly easy to collect, and may be stored for long periods of time without spoiling. Little wonder that the cereals—so named from Ceres, the Roman goddess of crops—have become the chief crop of most of the world's people. More than 70 percent of farmland is planted to cereals, which provide the human population with more than 50 percent of its calories.

Wheat and barley formed the basis for early civilization in the Near East; rice allowed the development of high cultures in the Far East; and maize* was responsible for the evolution of the great civilizations of the Americas. These four grasses, along with oats and rye, are the cereals best known to people of the Western world. In some parts of the world, particularly in Africa and Asia, other grasses, better adapted to climates too severe for the major cereals, are grown. The most important of these is grain sorghum. A number of different grasses, sometimes lumped under the name "millets," are also fairly extensively grown as cereals, as is Job's tears, perhaps best known in the United States in the form of beads used for necklaces. All of the grasses mentioned thus far, with the exception of maize, are native to the Old World, and it is sometimes stated that maize was the only true cereal ever domesticated in the New World. This is not quite true, for in prehistoric times Indians in southern South America cultivated a brome grass for their grain and in parts of Mexico a species of panic grass was grown. But neither of these was to achieve any importance. Not all the nutrition derived from grasses is in the form of grain—sugar cane, a grass, is the source of 50 percent or more of the sugar produced in the world—and quite apart from their use as food, grasses serve humankind in many other ways.

With the coming of civilization, our ancestors needed something to bring relief from the worries and cares that accompanied it. For this they often turned to the same plants that had played a fundamental part in making civilization possible. Fermented beverages can be made from a great variety of plants, and before civilization appeared people had probably already discovered that certain grasses serve to make excellent beer. It has even been suggested that the cereals were first cultivated in order to have grain available for making beer. Barley very early came to prominence in this regard, a position it has never relinquished. Today, of course, the grasses provide major ingredients for all of the popular alcoholic beverages except wine—rice is used in making *sake*, sugar cane for rum, corn for bourbon, rye and wheat for other whiskeys. Whether this use of grasses has been a service or disservice to humankind is a subject of divided opinions.

*Maize is the name of the plant that is commonly called *corn* in the United States. In other English-speaking countries, *corn* generally means the most common cereal, or some- times any cereal. In the Bible *corn* usually refers to wheat, as it does in much of the United Kingdom today—except in Scotland, where it refers to oats.

Figure 5–2 Bamboo in Indonesia. (Courtesy of FAO.)

Tropical grasses of tremendous importance are the bamboos, exceptional in the grass family for their size and woodiness. Practically indispensable in parts of the tropical world, they are little appreciated in temperate areas, although some people may recall that they were once used for the pole vault and they are still used for fishing rods. I know of no better way to tell of the importance of the bamboos than to quote from a paper that a Thai student, Sa-korn Trinandwan, once wrote for my course in economic botany.

I write this paper with gratitude and great respect for bamboo. In the poor families of the tropical people, the child is born on the floor of bamboo ribs under the bamboo roof in the bamboo hut. He is rocked in the bamboo cradle by his mother. He plays with toys which are made of bamboo. When

Figure 5–3 Some products of bamboo in Thailand. (Courtesy of Sa-korn Trinandwan.)

he grows up and does something wrong, he will be punished by his mother with a bamboo rod. Sometimes he entertains himself with the bamboo flute. Sometimes he eats the bamboo shoots which are cooked on the bamboo fire. Then he grows up to be a man and makes love under the bamboo shade and builds a bamboo hut for his wife. When he grows old and dies, his body is buried in the bamboo coffin. The bamboos give sadness and happiness to the life of man from generation to generation. Bamboos are parts of men's lives. Bamboos are a part of their blood, and they seek their ways to the human soul.

Bamboos can get along very well among the rich people. They get into the rich people's houses and serve them in the form of beautiful furniture. Sometimes bamboos get into their gardens and spend their entire lives with them. In the ancient times, even the king had to spend the night in the bamboo camp. During World War II, the prisoners of war spent part of their lives in the Japanese bamboo camps. How much the bamboos serve the man, whether rich or poor, high or low.

The importance of grasses for the feeding of livestock is fairly obvious. Although the saying that "all meat is grass" is not true, it comes close to being so since the grain and fresh or dried leaves of grasses are the principal foods of most of our domesticated animals.

Not to be overlooked is the importance of grasses as ornamentals. Although a few grasses are occasionally grown singly for their aesthetic value, the chief claim to significance of grasses as ornamentals comes in their use for lawns. With the widespread movement to suburbia in the United States in recent years, billions of dollars have been spent for blue grass seed or sod, for other lawn grasses, and for fertilizers, mowers, and so on. At the same time, a tremendous amount of money has been spent trying to eliminate undesirable plants from lawns, chief among them another member of the same family, crab grass.

In addition to crab grass, the family furnishes us with a number of other weeds, but this is equally true of most large plant families. There is no good scientific definition of a weed. Weeds are plants that follow people, growing best in the areas disturbed by them or their domesticated animals. They are sometimes referred to as "unwanted plants" or as "plants whose virtues have not yet been discovered." The second definition contains more than a modicum of truth. Rye and oats in prehistoric times were weeds of the cultivated wheat and barley fields. They spread with the domesticated plants as the latter were brought into new areas; in some regions they grew better than the wheat and barley, and in time they became intentionally cultivated and eventually domesticated.

Finally, in this brief introduction to the significance of grasses, mention must be made of their importance to soil conservation. A cover of grass affords greater protection against erosion of soil than any other plant, for the blades bend to cover the ground while rain is falling, forming a mat that permits the water to run off with very little or no loss of soil. Various grasses are thus often planted in areas subject to soil erosion. Some grasses too, particularly rye, are planted to be plowed under, to improve the texture and fertility of the soil.

People have used grasses in many other ways. But rather than explore these, let us consider at greater length the grasses that have contributed the most to our food.

Wheat

The most widely cultivated plant in the world today, wheat was one of the first two cultivated plants. The other, barley, has continued to be of importance, but chiefly as animal feed and as the source of malt for making beer. Wheat has become the principal cereal, being more widely used for the making of bread than any other cereal because of the quality and quantity of its characteristic protein, gluten. As it is gluten that makes bread dough stick together and gives it the ability to retain gas, the higher the proportion of gluten in the flour, the better it is for making leavened bread.

Bread, of course, was not one of the first prepared foods, for the early wheats were hardly suited to bread making. It seems likely that wild wheat and the early cultivated wheats were prepared by parching, which would have the advantage of removing the chaffy bracts surrounding the grain as well as making it more readily digestible. By grinding the parched grain and adding water, people could make a gruel. A beer of sorts may have been another early product in which grain was used. To make beer it would be necessary for some of the starch of the grain to be converted into sugar,* which then could be fermented by wild yeasts unknowingly introduced along with the plant materials. Long ago people found such a brew to their liking, and although they may not have known it, it was a nutritious drink. Yeast, in fact, may have become a cultivated plant before the grains did, for yeast-containing residues from beer made from wild ingredients might have been used for starting new batches, meaning that the yeast was actually being cultivated. Sometime later, leavened bread resulted from the same process, the carbon dioxide bubbles formed during fermentation becoming trapped in the sticky dough and causing it to rise. Unleavened bread, of course, was probably used earlier and continues to have significance in certain religious ceremonies.

Wild wheats, which still are found in the Near East in some abundance, were being collected by humans long before domestication occurred. Flint blades, which were apparently used for sickles to harvest grain, have been found in archaeological deposits dating back some twelve thousand years. Milling stones and querns for grinding are even older, although we do not know for certain that they were used for cereals. Evidence that the wild cereals could have supplied an abundance of food was provided by Jack Harlan, an American botanist, a few years ago. With a flint sickle he was able to harvest four pounds of wild grain in an hour. Thus, as he points out, a person equipped with such a tool could have in a space of a few weeks harvested more than enough grain to feed a family for a year.

---

*Starch is hydrolyzed to a fermentable sugar by the action of certain enzymes. Such enzymes could have been supplied in beer making by certain molds, by malting (that is, by germinating seeds), or by mastication. Malting barley is known to be quite ancient, being recorded in early written documents.

We may never know for certain why grain became cultivated, but we do now have considerable information about how wild wheat became transformed into the most important domesticated plant. We shall see that the process, which took place in the prehistoric period, involved accidental or natural hybridizations followed by a doubling of chromosomes. The working out of the details of the origin of domesticated wheats is as fascinating as any detective story (at least to the botanist) and the understanding of the origin of these plants has furnished knowledge of great importance to plant breeders in their efforts to improve them.

Our present understanding of wheat's origins is not the work of one or a few people but results from a great many botanical studies conducted during this century in Germany, Russia, Japan, the United States, England, and Israel. Taxonomists had described many different species of wheat during the previous 150 years, and Linnaeus had provided the genus name, choosing *Triticum*, an old Latin name for cereal. In the early part of this century, taxonomists recognized on the basis of the plants' appearance that there were three groups of wheat. Shortly thereafter it was shown that the three different groups were characterized by different chromosome numbers, the diploids having fourteen chromosomes, the tetraploids having twenty-eight, and the hexaploids forty-two. Thus, the wheat species form a polyploid series with a base chromosome number of seven.* At one time, many different species were recognized. Now the genus *Triticum* is believed to consist of only four or five species.

The species with fourteen chromosomes, *Triticum monococcum*, includes wild einkorn, which still grows wild in the Near East, and the cultivated einkorn, which differs primarily from its wild counterpart in having less brittle or nonshattering fruit stalks that prevent the grains

---

*Chromosomes, found in the cells of most organisms, contain the genes that control transmission of hereditary characteristics. The eggs and sperms of an organism each contain one set of chromosomes and are called haploid. The union of sperm and egg as a result of fertilization gives rise to a plant or animal, designated as diploid, that has two sets of chromosomes, one derived from each parent. Most plants and animals are diploid, but, particularly in plants, we may find individuals or species characterized by having more than two sets of chromosomes; these are known as polyploids. Polyploidy results from an "accident" in the chromosome division in a species or, more frequently, in a hybrid between species. In the wheats there are two sorts of polyploids: tetraploids having four sets of chromosomes and hexaploids having six. Chromosomes of a plant are studied by observation of dividing cells. Usually anthers in which young pollen grains are developing or young roots are used. The structures are crushed and treated with a stain that makes the chromosomes visible and then examined under a compound microscope at high magnification.

from falling readily from the plant. Einkorn, a low-yielding species, still occurs as a relic crop in Turkey, Caucasia, and parts of Europe. The chromosome sets of the diploid wheat are designated AA, each letter representing one set of chromosomes (Figure 5-4).

The twenty-eight chromosome wheats comprise two species, one of which, *Triticum timopheevii,* is only of minor importance. The other is *Triticum turgidum,* which includes a naturally occurring race, wild emmer of the Near East, and several domesticated races. The domesticated forms again differ from the wild type in having nonbrittle fruit stalks but both the wild and cultivated emmer have covered, or hulled, grains, a characteristic of wild grasses. The first of these wheats to appear was emmer which became important in the early period and was used both for making bread and beer. Emmer became more common than einkorn and spread in all directions from the Near East. Today emmer is still grown in the Near East, the Balkan countries, northeastern India, and Ethiopia. The other tetraploid wheats have naked grains which thresh free from the surrounding bracts, a boon to people, of course, but detrimental to a wild species. Among these is durum, or macaroni, wheat which is still very widely grown. The chromosome composition of the tetraploids is designated as AABB. Several hexaploid wheats, with forty-two chromosomes, are also known. These include varieties with hulled grains such as spelt, once the principal wheat of Europe but rare today, and some with naked grains such as bread, or common, wheat, now the most widely grown type and preferred for making bread.

In a polyploid series, the most ancient species is the one with the lowest chromosome number. From this it would follow that einkorn (AA) is the ancestor of the other wheats. Einkorn hybridized with another species with chromosome set BB and gave rise by chromosome doubling to the tetraploid, or AABB species. One of these tetraploids in turn hybridized with still another species, with the DD chromosome constitution, and this triploid hybrid in turn doubled its chromosomes to give rise to the hexaploid, or AABBDD, wheats (Figure 5–4). Attempts to identify the donors of the BB and DD chromosome sets have been the subject of intensive study over the years. Both of them come from goat grasses, weedy grasses worthless to humans except for their contribution of desirable chromosomes to wheat. The DD set, which contributes to the high gluten content of the hexaploids and also makes them better adapted to extreme environments than are the other wheats, comes from *Aegilops squarrosa,* a species that ranges from Turkey to Kashmir and Pakistan. Although various species have been suggested, there is yet no agreement on the contributor of the BB chromosome set. It is, of course, possible that the donor species is now extinct

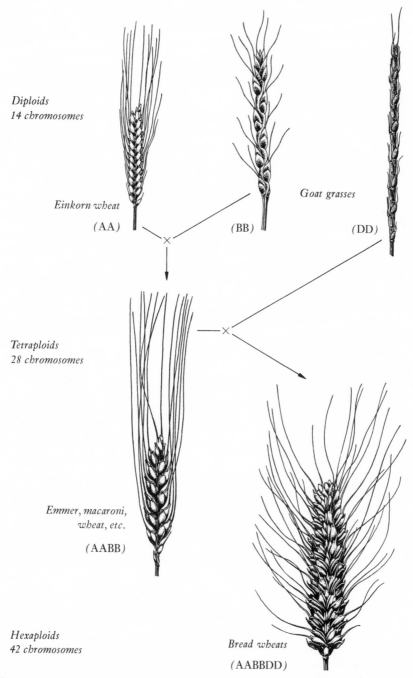

Diploids
14 chromosomes

Einkorn wheat

(AA)

Goat grasses

(BB)

(DD)

Tetraploids
28 chromosomes

Emmer, macaroni,
wheat, etc.

(AABB)

Hexaploids
42 chromosomes

Bread wheats

(AABBDD)

Figure 5–4 Evolution of domesticated wheats. (Modified from "Wheat" by Paul C. Mangelsdorf. Copyright © 1953 by Scientific American, Inc. All rights reserved.)

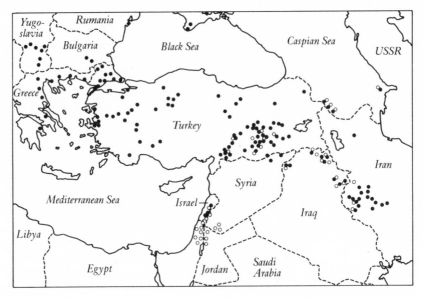

Figure 5–5 Distribution of wild einkorn (*solid dots*) and wild emmer (*open circles*). (Adapted from "Distribution of Wild Wheats and Barley" by Jack R. Harlan and Daniel Zohary, *Science* 153 (1966): 1074–80. Copyright © 1966 by the American Association for the Advancement of Science.)

or has changed to such an extent that it can no longer be clearly identified.

Archaeological work has helped to supply approximate dates for the development of the wheat species and their early dispersal out of the Near East. Wild einkorn has turned up in archaeological sites in the tenth and ninth millennia B.C. in Syria. Domesticated einkorn was present in the seventh millennium B.C. and went to Europe early with some of the other Near Eastern crops, appearing in Greece in the sixth millennium B.C. Wild and domesticated emmer have been found in several sites in the Near East in the eighth and seventh millennia B.C. Emmer soon reached Europe and became the dominant crop in ancient Egypt.* Free-threshing tetraploid wheats appeared soon after the establishment of emmer in the seventh millennium.

---

*Occasionally an item is published in a newspaper to the effect that Mr. and Mrs. So-and-So on a vacation to Egypt visited the pyramids and brought home with them a few grains of "mummy" wheat, several thousand years old, and that the grains were planted and gave rise to healthy wheat plants. While it is true that grains of wheat dating back to pre-Christian times have been found in Egypt, there is no scientific verification of their ever germinating. Wheat seeds, like the seeds of most plants, have a rather short life, lasting only a few years.

Spelt, the first hexaploid, became known in Transcaucasia early in the fifth millennium B.C. after the tetraploids had spread north into the range of *Aegilops squarrosa*. The acquisition of the chromosome set from this goat grass increased the ecological amplitude of wheat and helped the wheat spread in the areas of Europe and Asia with continental winters and humid summers. It is sometimes difficult to distinguish the grains of the free-threshing tetraploids from the free-threshing hexaploids, although the latter generally have plumper grains. However, free-threshing hexaploids probably came sometime in the fifth millenium B.C. The free-threshing hexaploids moved into many areas, one of them, bread wheat, later becoming the dominant type around the world. Hexaploid wheats reached India before the fourth millennium B.C. and reached China sometime before the Christian era. The Spanish brought wheat to Mexico in 1529, and eventually the United States, Canada, and Argentina were to be among the world's greatest producers. (As an indication of the way technological and scientific development is drastically altering agricultural practice, it is interesting to note that the United States, where wheat is a relatively new crop, has sent agricultural specialists to give advice on growing wheat in Turkey, where the crop has been grown for thousands of years.)

Wheat grows best in cool climates with little to moderate amounts of rain, and many areas of the earth's surface are suitable for its cultivation. Wheat is being harvested somewhere in the world every month of the year. Of the several kinds of wheat brought into domestication, only two are of much importance today: bread wheat (*Triticum aestivum*) which furnishes about 90 percent of the wheat grown, used primarily for flour for bread and pastries, and durum (*Triticum turgidum*), employed principally for making paste products such as macaroni, spaghetti, and noodles. Both species are cultivated in the United States, the latter principally in North Dakota. In the northern part of the United States and in Canada, wheat is planted in the spring; to the south of this area it is sown in the fall.

Although in some parts of the world the methods employed in growing, harvesting, and utilizing wheat are only little changed from those used in prehistoric times, in other places there have been great advances in wheat farming and in processing the grain—from the introduction of modern mechanical planting and harvesting to steam-driven rollers for milling. In the first part of this century it still took tremendous human effort to thresh and harvest wheat, even in the industrialized countries, although some machinery was used. The combines used today consolidate what were formerly many separate operations of threshing and harvesting and can be operated by a single person. The combine has

A

B

C

D

Figure 5–6 Traditional and modern methods of harvesting wheat. *A.* Harvesting wheat in India. (Courtesy of FAO.) *B.* Threshing wheat with oxen in India. (Courtesy of Rockefeller Foundation.) *C.* Milling wheat in Pakistan. (Courtesy of FAO.) *D.* Combining wheat in the United States. Air-conditioned combines are now widely used. (Courtesy of USDA.)

thus led to drastic changes in farm life, including the virtual disappearance of the old "threshing party." Fifty years ago straw stacks were a common sight on farms in the midwestern United States—now the straw is usually scattered on the fields in the combining process, or neatly baled. Wheat straw, although of minor importance compared to the grain, has some uses. It is not a nutritious hay for livestock but is unusually strong and has many uses on a farm; it can also be used by manufacturers as a filling for mattresses and in the making of paper and paper board.

Bread, too, has changed. To make a white flour, it is necessary to remove the germ and the bran from the grain. Technological advances in the milling process over the years have led to a whiter flour. Since the natural product is somewhat yellowish in color, a bleach is sometimes employed, and chemicals such as sodium propionate are sometimes added during the baking process to retard spoilage of the bread. Although white flours have superior keeping and baking qualities, they are less nutritious than the whole wheat. Therefore, the flour is often "enriched" by the addition of vitamins and minerals, which is required by law in some countries. These, however, fail to compensate for all the nutrients and fibers that have been removed from the grain. In the past a white bread was associated with the higher class or at least those with money. Recently there has been a trend toward whole-grain bread among some of the better educated. It is a sad note on modern civilization that we must remove some of the nutrients and then supply additives to the product that goes on the table.

There have been changes in the wheat plant itself, as well as in methods of growing, harvesting, and milling it. Much of this evolution is the result of human selection and began when people first started growing the wild grasses. As the wild plant had a brittle fruiting stalk, the grains tended to fall individually as they matured, an advantage for seed dispersal in nature but undesirable for humans, who wanted to collect all the grains at one time. With domestication the fruiting stalk became less brittle, permitting the grains to remain on the plant. The cultivated plants developed seeds that would germinate rapidly and evenly, in contrast to the slower, more irregular germination of the seeds of the wild plant. The bracts surrounding the grain became more loosely attached, letting the grains fall free of the bracts during the harvest, which made it easier to prepare them for eating. There was also some increase in the size of the grain. Numerous varieties of the different species came into existence through the agencies of mutation, hybridization, and both conscious and unconscious selection. Possibilities for hybridization increased as people moved from place to place or exchanged seeds with others.

As new varieties of wheat come into existence, some of the older ones tend to disappear, but many have existed little changed for hundreds of years. In this century plant breeders, through conscious selection and artificial hybridizations, have produced so many new varieties that there are now probably more than a thousand different kinds of bread wheat alone, some of them tailor-made for high productivity, special milling properties, adaptation to different climatic conditions, and resistance to disease.

Perhaps it is resistance to disease that has occupied more of the plant breeder's attention than any other characteristic. Practically all of the cultivated plants are subject to a number of diseases, and the wheat plant is particularly susceptible to stem rust, caused by a fungus that belongs to the same group as the common mushroom but is profoundly different in appearance. A severe infestation of stem rust can almost completely destroy a crop. Many races of wheat have been bred that are resistant to rusts, but the great difficulty is that there are also numerous races of rusts and new, more virulent ones continually come into existence through mutation and hybridization. Thus there will probably always be a need for the plant breeder to attempt to develop new resistant strains of wheat. Another accomplishment of plant breeders has been their contribution to the doubling of the wheat yield in the United States in the course of little more than a quarter of a century. Greater use of fertilizers and improved methods of preparing soil and eliminating weeds have also contributed to the increased yields.

Among the recent achievements of the wheat-plant breeder and other agricultural scientists is the remarkable work accomplished in Mexico that contributed significantly to the Green Revolution. In 1943 members of the Rockefeller Foundation, at the invitation of the Mexican government, went to that country to see what could be done to increase food production. Although maize is the basic food plant of Mexico, wheat is also very important and Mexico was then importing half of the wheat it consumed. Much of the wheat being grown in Mexico at the time was little changed from that originally introduced by the Spanish in the early part of the sixteenth century. Working with Mexican agronomists, the scientists from the Rockefeller Foundation were able to double the yields in twenty years. A large part of the success resulted from the development through hybridization of rust-resistant varieties and semidwarf forms able to take heavy applications of fertilizer without falling over. The older, taller varieties tended to lodge, or fall over, when given sufficient fertilizer to increase the yields and thus could not be readily harvested. The parental type of the new semidwarf forms that have proved so successful originally came from the Orient. The first crop planted by Norman E. Borlaug, who headed the work for the Rockefeller

Foundation and in 1970 received the Nobel Prize for his contributions to agriculture, was lost to rust. Fortunately, a few of the original seeds were still left in the seed envelopes, and these were grown the next year and used for the crosses that ultimately gave the kinds of plants that were wanted. Some of the wheat varieties developed for Mexico have proved to be successful in other parts of the world. Pakistan is now producing enough wheat for its own needs, while India is approaching that goal. It is reported that in parts of India that used to produce seven or eight bushels an acre, yields of up to sixty or seventy bushels an acre are being secured by growing the new varieties under irrigation. In India the new high-yielding varieties originally attracted such attention that armed guards had to be posted at the experimental stations to prevent people from stealing seed before it was ready to be released to the public. Since wheat is a self-pollinated crop, once a hybrid has been stabilized and released to farmers they can multiply their own seed stock for future planting.

Although it is always important to produce a plant that can yield more per acre, production of wheat was so great in the United States in the early 1960s that it included what has been referred to as an "embarrassing" surplus, more than could be utilized at home or sold for export. It has been easier to breed high-yielding wheats than it has been to solve the problems of their distribution.

A hybrid between wheat and rye has been known since the last cen-

Figure 5–7 Dr. Norman E. Borlaug (*fourth from left*) showing one of the new wheats developed in Mexico to a group of visiting scientists. (Courtesy of Rockefeller Foundation.)

Figure 5–8 Armed guard protecting plot of a new strain of wheat at an experimental station in India. It was necessary to post guards so that the farmers would not take the seeds before they were ready to be released to them. (Courtesy of Rockefeller Foundation.)

tury. In 1935 attempts were begun in Sweden to develop it as a crop plant. Although most of the early hybrids were sterile or nearly so, plant breeders developed methods to produce fertile types. The hybrid, popularly known as triticale and known scientifically as *Triticosecale* (from *Triticum*, wheat, and *Secale*, rye), has a nutritional quality similar to that of wheat but superior in its higher lysine content. Although not as suitable as wheat for making bread, it is as good as wheat for most other products. Yields are now reported to be equal to that of wheat and the hybrid will grow in some areas ill-adapted for wheat. In 1986 a total of about two and a half million acres was grown in thirty-two countries. France led the world in production followed by Poland and the Soviet Union. Research is continuing on this cereal, particularly in Mexico and Europe, and we can expect it to make more significant contributions to our food supply in the future.

Rice

In Japan each year the emperor himself, still patron of all agriculture although no longer regarded as the descendant of the sun, joins in the

Figure 5–9   Rice. (Couresy of FAO.)

ritual harvest of the rice on the small imperial paddy in the palace grounds. Rice is still regarded as a sacred plant by many people in much of Asia. Although of more limited distribution than wheat, rice feeds more people, since it constitutes the basic food of more than half the world's population.

The genus *Oryza* comprises twenty species found in the humid tropics of the Americas, Africa, Australia, and Asia. One species, *Oryza glaberrima*, was domesticated in West Africa, where it is still cultivated, and another species, *Oryza sativa*, which furnishes virtually all the rice of the world today, was domesticated in Asia. The wild ancestral species of the latter is generally given as *Oryza rufipogon*, which is still widespread in Asia and in some places hybridizes with cultivated rice. There is some recent evidence that domestication occurred in two places, China and southern Asia. With selection, annual plants with larger grains, non-shattering fruit stalks, and sometimes lacking the awns of the wild species were developed. The archaeological record of rice is far inferior to that of wheat, but present evidence indicates that it was in cultivation in both India and China before 5000 B.C. Wild rice is still gathered by some peoples and is the preferred plant in some religious ceremonies in southeastern Asia.

Rice reached Japan from China during the second century B.C. At this time it was already known in Europe, for it had been carried to Greece by Arab traders and again during Alexander the Great's invasion of the East. The scientific name *Oryza*, applied to the plant by Linnaeus, comes from an ancient Greek word for rice that in turn is derived from either an Arabic word or a Chinese word—opinion is divided among scholars. The Chinese word means "good grain of life"; not surprisingly, the word for rice is the same as that for life, food, or agriculture in many parts of the Far East. Rice is mentioned in very ancient Chinese writings, as it is in early Hindu scriptures. Several different varieties of rice are described in the ancient Dravidian literature of India. Linguistic evidence, as well as botanical, suggests an origin for the plant somewhere in southeastern Asia.

Rice was introduced into the Carolinas in 1647, but today California, Arkansas, Louisiana, and Texas are the rice-growing areas of the United States; more than half of the rice produced is exported. As the consumption of rice in the United States is only six pounds per person per year, not a great deal need be produced to meet the domestic demand. Some rice is also grown in Africa, South America, and Europe. In many parts of Latin America, rice and beans are served with every meal, and sometimes they constitute the whole meal. But it is in its original homeland that rice is dominant, where more than 90 percent of the world's crop is produced, and where the average person consumes as much rice in two weeks as the average American eats in a whole year, although rice consumption has doubled in the United States in the last twenty years.

Throughout much of southeastern Asia, rice is grown as it has been for many centuries, requiring tremendous human effort. It has been estimated that in some regions it takes one person one thousand hours to grow and harvest a single acre of rice. A great many people still cultivate very small paddies—from one to five acres—of rice.

As methods vary somewhat from region to region, a generalized description is impossible, but the following steps are very common. The fields are prepared for planting with a wooden single-furrow plow drawn by a water buffalo or ox. Manure, if available, is scattered on the field. The plowed land is then smoothed with a log. After the dikes have

Figure 5–10 Terraced paddies in Indonesia. Mechanization would be difficult in this kind of terrain. (Courtesy of FAO.)

been repaired, river water drawn by primitive water wheels is used to flood the fields in areas where rain is not sufficient for this purpose.

The rice is then planted, either by broadcasting dry or previously germinated seed or by transplanting seedlings or young plants that have been grown in a nursery bed. The customary way of planting most cereals is directly by seed, and knowledge of the origin of transplanting seedlings by hand would be of interest. Carl Sauer has postulated that the latter is the older method, first used by people who had practiced vegetative cultivation of noncereal crops, transplanting being similar to starting plants from cuttings or by other vegetative means. Broadcasting dry seeds might have been derived from contact with wheat farmers at a later date. It is not definitely known which is the older method. Transplanting, backbreaking work that requires much stooping, is usually

A

B

Figure 5–11 *A.* Farmer hoeing a paddy in Indonesia. Although water buffalo (Figure 4–11) are widely used, much hand labor still goes into the preparation and care of the fields. (Courtesy of FAO.) *B.* Transplanting rice seedlings in Japan. (Courtesy of FAO).

done by women and children. Some pruning of the tops and the roots of plants normally accompanies transplantation, and some people believe that this pruning stimulates growth of the young plant and leads to higher yields than when seeds are broadcast. Weeds present less of a problem when rice is transplanted than when seed is planted directly, and having less competition from weeds could also help account for greater yields being obtained through the transplanting method. After the plants are established, any necessary weeding is done by hand, another activity that requires much stooping.

Harvesting is done with sickles or knives. With the latter the seed heads are gently cut individually, for the idea still exists in some regions

Figure 5–12   Harvesting rice in Indonesia. (Courtesy of FAO.)

Figure 5-13   Winnowing rice in Burma. (Courtesy of FAO.)

that if the cutting is harsh, the rice plant will be offended and the yield will be decreased the following season. Threshing is done by beating the heads against the ground or against logs, or by having animals or barefoot humans tread upon the seed heads. Sometimes women who do the treading are barebreasted, which is thought to be related to an ancient belief that the less they wear, the thinner the rice husks will be. Winnowing to remove the chaff from the grain is still often accomplished by tossing the rice from bamboo or rattan trays and allowing the wind to blow away the lighter chaff while the grain settles nearby. All of these methods, of course, contrast strikingly with the highly mechanized procedures practiced in the United States and some other countries, where airplanes are sometimes employed for seeding, allowing the grain to be produced economically in spite of high labor costs.

Following winnowing, under primitive conditions, a mortar and pestle are used for hulling and the brown rice that results may be promptly cooked for local consumption. In industrial preparation pearling or whitening removes the bran or outer layers of the grain, and sometimes polishing follows. The final product is pleasing in appearance and taste but less nutritious than brown rice. Unfortunately, just as with bread, most people prefer white rice to brown. The loss of nutrients, particularly vitamin B_1 eliminated in the process of milling and cooking, has been responsible for deficiency diseases such as beriberi among people whose diet consists almost entirely of white rice. Highly milled rice contains only 0.04 milligrams of vitamin B_1 per 100 grams, compared to 0.4 milligrams in unmilled rice; the latter amount is sufficient to prevent beriberi in someone who eats rice daily. White rice, of course, can be fortified with additives to increase its nutritional value.

Rice, although not a true aquatic plant, is unusual among the cereals in that its roots can thrive under water. Thus it is an ideal plant for much of the humid tropics, although at times it can get too much water, for proper drainage is essential to its good growth. The presence of a weed, the water fern, *Azolla*, in rice paddies has long been known to stimulate the growth of rice. It was found that the water fern is inhabited by the blue-green alga, *Anabaena*, which supplies nitrogen to the rice through its ability to fix atmospheric nitrogen (see Chapter 7). In some parts of Asia the water fern is deliberately planted in rice fields, thus eliminating the need for expensive nitrogenous fertilizer. Rice also can be grown much as the other cereals. Dry upland rice, or hill paddy, is grown in some areas that have the proper temperature and sufficient rainfall. Yields, however, are generally lower than for lowland rice. In some parts of Asia two crops of lowland rice can be grown in a year.

The thousands of varieties of rice are divided into three subspecies, two of which are widespread: the *japonica* types, which have short grains and are sticky when cooked, and the *indica* types, which have long grains and are drier when cooked. The *japonica* types, grown in both Japan and Taiwan, in general are higher yielding than the *indica* types, which are grown throughout the greater part of southeastern Asia. World hunger could be at least partially reduced through the wider distribution of the higher yielding varieties. One of the obstacles to achieving this goal, however, lies in the difficulty of persuading people to accept new foods. People the world over usually prefer their local staples, and if they are used to eating a sticky rice, for example, they are often reluctant to change to a dry type, and vice versa.

The rice plant, directly or indirectly, serves humankind in many ways other than for food. Beers and wines may be made from the grain, the most famous of which is *sake*, the "national beverage" of Japan. Rice is also used in the manufacture of beer in the United States and other countries. Rice is used for the production of starch, and rice powder is used as a cosmetic in parts of the Far East. The hulls, or husks, of the grain are used as fuel, in making building materials, for the manufacture of the chemical furfural, which in turn is used to make plastics, and in other ways. The straw, of course, is not overlooked and is used in the manufacture of baskets, mats, and strawboard. Like wheat straw, rice straw is not very nutritious, but it is sometimes used for livestock feed. Paper can be made from rice straw, but what is commonly known as rice paper is made from another plant—*Tetrapanax papyriferus,* or rice-paper plant—which is native to the Orient but is not a grass.

Perhaps the most important by-product of rice farming doesn't come from the plant itself, for paddies are frequently used for raising fish. A rice diet is extremely deficient in protein, and fish, of course, is an excellent protein source. In many rice-growing areas ponds are kept for fish hatcheries, and small fish, usually carp, are introduced into the paddies. The fish, rather than decreasing rice production, actually appear to promote its fertility. Without the fish in the wet rice fields, other problems may arise—as was discovered in the south coastal area of the United States some years ago when huge swarms of mosquitoes found the rice fields ideal breeding grounds.

Of the many diseases of rice, one has proven to be of particular scientific interest. The *bakanae* or "foolish seedling" disease, which produces unusually tall, thin plants, was found by Japanese botanists to be caused by a fungus, *Gibberella fujikuroi.* They found that a growth substance, now known as gibberellin, could be isolated from the fungus, and this substance has been the subject of considerable scientific work by American and English botanists as well as the Japanese since World War II. Gibberellins are used to produce growth in some dwarf plants and to induce flowering in others.

Although some efforts to improve the rice plant have been carried on for many years, and high-yielding strains have been grown in Japan and Formosa, much of Asia still grows its old unproductive types under very primitive conditions. Recognizing the need for improvement in a plant that feeds a great part of the world's population, the Ford Foundation, in cooperation with the Rockefeller Foundation, established the International Rice Research Institute (IRRI) in the Philippines. With scientists representing many different disciplines assembled from the United States and six Asian nations, the IRRI began its work in 1962. It was soon found that, although rice was a much studied plant, there was still much room for basic research. Effort was devoted to learning more about the plant, and this search for basic knowledge still continues. One of the institute's greatest services has been to assemble the largest rice collection in the world—over sixty-seven thousand varieties, mostly from Asia, some of which were on the verge of extinction. Many of these are of no significance today, but they may furnish valuable genes for the improvement of rice in the future.

It was already known that one of the great drawbacks of rice was its tendency to lodge, or fall over, as the plants reached maturity. Fertilization, which would improve yields of such plants, would only contribute to lodging since it would produce taller plants with heavier heads. The rice scientists realized that an ideal rice plant would be a short one with a strong stem that could take additional amounts of fertilizer. Over ten thousand different samples of rice seed from many different areas were assembled, and through hybridization the IRRI scientists attempted to produce a plant having the desirable characteristics. One of the crosses, involving a short rice from Taiwan called Deegeowoogen and one from Indonesia called Peta, gave extremely high yields not only in the Far East but in South America and Africa as well. The new plant, designated IR8, reaches maturity in 130 days and is insensitive to day length, making it adaptable to many regions and making it possible to grow more than one crop a year in some places. Other high-yielding varieties have since been developed. The IRRI did not neglect other aspects of growing rice—fertilization, irrigation, disease control—and even produced an inexpensive threshing machine that could be widely used.

Thus, in the space of a few years, through a highly concentrated effort, rice yields increased dramatically and contributed greatly to the Green Revolution. The newer rices have also had some disadvantages; for example, the greater use of pesticides to secure the higher yields has had deleterious effects on the fish populations in the paddies. The most significant recent improvement has been the development of hybrid rice in China. By 1986 nearly twenty million acres of it were being grown there.

Wild Rice

Wild rice has been mentioned previously as the ancestor of rice, but there is another plant also known as wild rice. This wild rice, *Zizania palustris*, is a grass, but other than that it is not closely related to true rice, *Oryza*. Wild rice is native to the Great Lakes region, where it grows in wet habitats similar to those of true rice—hence the common name. Wild rice was an important food source to the native North Americans particularly in Minnesota, Wisconsin, and Manitoba. Until recently it was still collected by them for their own use and as a cash crop. Twenty years ago it became a cultivated plant in northern Minnesota and more recently in northern California. The wild grass has shattering grain, not a desirable character for a cultivated cereal. Plants that retained most of their grain were found to serve for cultivation, however. Fields are flooded for growing the plant and then drained for harvest, which is done with large combines. Wild rice yields only about fifty pounds an acre when harvested in the wild but this has greatly increased under cultivation, sometimes reaching a thousand pounds an acre. Wild rice is considered a delicacy among the grains and appears to have a bright future. It is also noteworthy because it is one of the very few wild food plants to have become domesticated in recent times.

Maize

From a plant that was once used mostly to feed people, maize has become one of the world's chief feeds for animals. Maize, however, still retains its role as a basic human food plant in many parts of the world, particularly in much of Latin America. It is generally held that maize was not known in the Old World until after some of Columbus' crew found it growing in Cuba. Maize was already several thousands of years old at the time, and it had fed the laborers who built magnificent temples in Mexico and Peru in prehistoric times; today, directly or indirectly, it supplies much of the energy for technological developments in the United States.

At the time of the discovery, maize was the most widely grown plant in the Americas, extending from southern Canada to southern South America, growing at sea level in some places and at elevations higher than eleven thousand feet in others. Woodlands were cleared, swamps drained, deserts irrigated, and terraces built on mountain sides so that the plant could be grown, but little maize grew in the area that was to become the great corn belt of the United States, for not until the mould board plow was invented would it be possible to turn the heavy prairie sod. For many years after America was discovered by Europeans, much of the future corn belt was to continue to be dominated by wild grasses and buffalo.

Much of Native American life centered on maize, the "gift of the gods," just as it does today in parts of Mexico and other Latin American countries. In religion and art, as well as in everyday life, many human activities were concerned with the maize plant. For food, it was popped—probably one of the oldest ways of preparing the hard grain—parched, boiled, or ground. By washing the grains in wood ashes and quicklime, maize was made into hominy. Ground, it was used for making unleavened "bread," or tortillas. It was also eaten green; from quids, or chewed wads, recovered in archaeological deposits, we know that this was one of the most ancient methods of eating it. Sometimes even the pollen was added to soups or stews, and corn smut, a fungus disease that affects corn, was also eaten. As Paul Weatherwax has written,

Figure 5–14 Native American methods of preparing bread and chicha. (From Jerónimo Benzoni, 1565.)

"Perhaps the most cheering and heart-warming use the Indians made of maize was the production of alcoholic beverages." It was probably learned early that a nutritious beverage could be made by chewing, or germinating, the grain to start a process of fermentation. The beer, or *chicha*, that results is still widely made and used in South America, although in many areas, unfortunately, it has been replaced by the much more potent and less nutritious sugar-cane alcoholic beverage, *aguardiente*. People in the United States learned that a whiskey could be made from corn and American bourbon was born.

In Europe maize was originally grown as a curiosity, like many other American plants of the time. To distinguish it from the other cereals, it was at first called Turkey corn, or Turkey wheat, because some people thought that it came from that country.* A Native American name, *maize*, or *mays*, was used to some extent. Linnaeus, the great Swedish botanist of the eighteenth century who named many of our plants, adopted it for the specific designation of the plant and used *zea*, an ancient Greek word for cereal, as the genus name. Thus we still have *Zea mays* as the scientific name of maize.

The new plant was not exactly a rousing success in all parts of Europe. John Gerard, the famous English herbalist of the late sixteenth and early seventeenth century, had this to say:

> Turky wheat doth nourish far lesse than either wheat, rie, barly, or otes. The bread which is made thereof is meanely white, without bran: it is hard and dry as Bisket is, and hath in it no clamminesse at all; for which cause it

---

*Some people have maintained that maize reached Asia before Columbus discovered America and that it may have entered Europe by way of Turkey.

Figure 5–15 One of the first published illustrations of maize in Europe, from herbal of Leonhart Fuchs, 1542. Fuchs believed that the plant came from Asia and called it Turkish corn.

is of hard digestion, and yeeldeth to the body little or no nourishment. Wee have as yet not certaine proofe or experience concerning the vertues of this kinde of corne; although the barbarous Indians, which know no better, are constrained to make a vertue of necessitie, and thinke it a good food: whereas we may easily judge, that it nourisheth but little, and is of hard and evill digestion, a more convenient food for swine than for man.

More than two centuries later, the English imported corn from America as food for the Irish people during the great famine. The Irish did not take readily to the new food. They did not know how to prepare it, they had no good implements for grinding it, and when they did, they didn't like the taste. Their inability to accept and use a New World plant is rather ironic, for it was the failure of the potato crop that led to the famine, and the potato had been introduced to Europe from the Americas even later than maize.

Indeed, the maize plant was a strange one to the Europeans—an amazing plant, if you will excuse an old pun—quite unlike the other cereals known to them. Not only was it larger but, instead of bearing its grains in a head at the top of the plant, it bore them in ears on the sides of the stalk, with the silks, or styles, protruding. At the top it produced tassels, which, except in exceptional circumstances, produced only male flowers. The grains, instead of being covered individually by chaff as in other cereals, were naked,* and the ear as a whole was covered by husks. Although nearly all domesticated plants are poorly equipped to survive in the wild, maize is even more helpless than most, because the grains remain attached to the cob. Maize had to be handled differently for planting than the cereals familiar to the Europeans. Instead of being broadcast over the field, maize grains, which are much larger than those of other cereals, were planted individually. This method of planting may relate to a basic difference in very early agriculture between the Old World and the New. In the Old World people had animals to plow the fields in preparation for the broadcasting of seed, whereas in the New World the only animals available were people themselves, and what crude preparation the field received was done with a digging stick or hoe. The maize farmer would have focused more attention on the individual plant, for which he had carefully sown a single seed, than did the wheat farmer who had broadcast seed; this may have been significant in the evolution of maize, for attention paid to variant individuals may help explain its great diversity.

Maize is certainly a most variable species, probably more so than any

---

*There is a rather rare form of maize, known as pod corn, in which the individual grains are covered.

Figure 5–16 Growing maize in Peru. *A*. Planting. A footplow is used to break soil. *B*. Cultivating. (From Poma de Ayala, c. A.D. 1600.) *C*. Preparing a field for planting today. Note similarity of plow to that shown in *A*. (Courtesy of FAO). *D*. Terraces. Many of the terraces constructed by the Incas are still in use at present. (Courtesy of Paul Weatherwax.)

Figure 5–17 A collection of maize from Guatemala showing some of the great variation found in this plant. (Courtesy of USDA.)

other plant. Many people are well aware of the great variation in the color of the grains, since multicolored ears are often used as decorations. The kernels also vary considerably in size and shape, as do the ears, which may range from a little over one inch to a foot and a half long. Some varieties are known that reach maturity in a little over two months, whereas others require more than a year. For convenience, five main types of maize are recognized: (1) popcorn, probably the most primitive type, with extremely hard grains that allow pressure to be built up within them upon heating; "popping" results when the sudden expansion of the soft starch turns the grain inside out (other varieties of maize and other cereals can also be popped and are commonly prepared in this way as breakfast foods, but special methods are necessary to allow the pressure to build up); (2) flint corns, which have kernels of hard starch; (3) flour corns, which have soft starch, of particular value to the Native Americans because it is easily ground, but disadvantageous in being quite subject to insect damage; (4) dent corns, so called because there is a dent in the top of the kernel in which a soft starch overlies an area of hard starch; modern dent corns are responsible for the high productivity of the corn belt and were originated in the nineteenth century by crossing a southern dent corn with a northern flint variety; and (5) sweet corn, which has sugary instead of starchy kernels and is today a favorite fresh vegetable. There is also another type of corn, called waxy corn, that is distinguished by a starch that is chemically different from that of other corns. It receives its name from the waxlike appearance of the grain when it is cut. Waxy corn was discovered in China at the beginning of this century and has come to have some special uses in industry and for food because of its very different kind of starch. In spite of this great variability, all maize is regarded by botanists as belonging to a single species. All the varieties are diploid, having ten pairs of chromosomes.

Where this species came from has long been a question of great interest to botanists, and at times proponents of different views have had rather heated arguments. Although much remains to be learned, there now appears to be some general agreement about where maize must have originated and the approximate time when this occurred, as well as about what happened to the plant under human influence. There also appears to be growing agreement about what plant gave rise to maize.

It has been known since the last century that teosinte, a coarse annual* wild grass of Mexico, Guatemala, and Honduras, now becoming

* A new perennial teosinte has been discovered in southern Jalisco, Mexico (H. H. Iltis et al., *Science* 203 (1979): 186–188), that may have important implications bearing on the origin of maize and may prove useful in breeding.

rather rare, was the closest relative of maize. The two plants cross naturally and produce fertile hybrids. Teosinte was originally considered to belong to a distinct genus and was classified as *Euchlaena mexicana*. The later recognition of its very close relationship with maize prompted its transfer to the genus *Zea*, and quite recently the proposal has been made that it belongs to the same species as maize. The idea that teosinte might be the ancestor of maize is not a new one, but it was largely ignored for a number of years in favor of the hypothesis that maize evolved from wild maize and that teosinte represented a descendant of maize or the hybrid of maize with some other plant. The hypothesis that wild maize was the progenitor of the domesticated plant has been championed by Paul C. Mangelsdorf since 1939. The hypothesis that teosinte was the direct ancestor was revived by George Beadle in 1972 and now has a growing number of adherents. On one point, however, there does seem to be general agreement: maize had its origin in Mexico.

Bushels of corn cobs and other parts of the plant have been recovered from archaeological deposits, mostly in Mexico and the United States. One of the most significant discoveries was made in Tehuacan, where cobs only little more than half an inch long (Figure 1-1) were found in deposits dated at between 5000 and 3400 B.C. Mangelsdorf postulated

Figure 5–18 Principal varieties of maize. *Left to right:* popcorn, sweet corn, flour corn, flint corn, dent corn, pod corn. (Courtesy of USDA.)

A

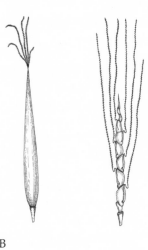

B

Figure 5–19   A. Teosinte. (Reprinted with permission of The Macmillan Company from Paul Weatherwax, *Indian Corn in Old America.* Copyright © 1954 by The Macmillan Company.)
B. Teosinte ear on left, about three inches long. Husks removed to show single row of grains on right. The grains are separate at maturity and fall individually.

that the cobs came from wild maize, but others consider them representatives of an early domesticated form of maize.

One of the difficulties in accepting teosinte as the direct ancestor has been to explain the great differences in the reproductive parts between maize and teosinte.* The two show more profound differences than those usually separating a wild progenitor from its derived domesticated form. It has been shown biochemically, however, that teosinte is as similar to maize as our most wild plants are to their domesticated counterparts. Another stumbling block in accepting teosinte as wild maize was the difficulty in understanding how people could have used its extremely hard grains for food. Nonetheless, Beadle showed that the

*Hugh Iltis has recently proposed (*Science* 222 (1983): 886–894) that the major change from teosinte to maize occurred as the result of a "catastrophic sexual transmutation."

grains could have been popped or ground to make a tortilla of sorts. Those rejecting teosinte as the ancestor have also pointed out that with one possible exception it has never been recovered from archaeological deposits. Neither has wild maize, however, if the material from Tehuacan is not accepted as such. Moreover, wild maize has never been found in the living state, so if such a plant ever existed it is probably now extinct. That people took such a seemingly unpromising plant as teosinte and made it into one of the world's greatest cereals may strain the belief of some, but this is the hypothesis that has now gained considerable acceptance.

From Mexico maize went to South America, reaching Peru in around 3000 b.c. It shows as much diversity in Peru as it does in Mexico, and at one time some botanists thought that it might have originated in Peru or been independently domesticated there as well as in Mexico, but those hypotheses have fallen into disfavor. Sometime later maize reached what is now the United States, and it was widely cultivated in both the Southwest and the East when the Europeans arrived.

Although presumably maize was a most mutable plant, it must have required considerable conscious selection to have produced the great number of varieties that existed when it was discovered by the Europeans. The development of a wild grass into the "happy monster" that once fed most of the people of the Americas was a most significant achievement of primitive plant breeders. The development of hybrid corn in this century ranks as one of the most outstanding developments of the modern plant breeder.

The story of hybrid corn has been told many times and in many places, but it bears retelling. First, some general remarks about hybridization are in order to set the stage. I have already mentioned hybridization several times in this book—both natural hybridization (between and within species) and artificial hybridization for plant and animal improvement. In either natural or artificial hybridization, genes may pass from one species or variety into another as a result of the back-crossing of the first-generation (F_1) hybrid to one of its parent types; people may exploit the new types that result. Such gene transfer may occur from many hybrids between species, but not from all of them by any means, for some interspecific hybrids are sterile or nearly so, like the mule. However, hybrids within a species, for example, between two different varieties, are generally fertile; this is true of hybrids between different varieties of maize. First-generation hybrids frequently show a greater vigor than either of their parents. This *hybrid vigor*, also called *heterosis*, is seen in the mule and was to prove to be one of the greatest significance in corn.

Two of our greatest biologists, Charles Darwin and Gregor Mendel, figure at least indirectly in the production of hybrid maize. Darwin in some of his research found that continual inbreeding of plants reduced vigor, whereas crossing different varieties increased vigor. Darwin even used maize in some of his experiments, although nothing more was to come of it at the time. Mendel supplied us with the laws of genetics which, after their rediscovery in 1900, made it possible to use deliberate planning in developing better plants and animals.

The history of hybrid corn begins with a man much influenced by Darwin, William James Beal, who in 1877 made the first controlled crosses in maize in an attempt to increase the yield. In his work at Michigan Agricultural College (now Michigan State University) he proved that yields could be increased by crossing different varieties. Shortly after the turn of the century two researchers began to study the effects of inbreeding: George Harrison Shull at Cold Spring Harbor, New York, who was following up some of Mendel's discoveries, and Edmund Murray East at the Connecticut Experiment Station. Their self-pollinated maize plants became weaker after each generation, but they found that if two inbred varieties were crossed, great vigor could be restored in this single step. The final touches for the ultimate commercial production of hybrid corn were provided by D. F. Jones, a student of East at Harvard, who went to the Connecticut Experiment Station in 1914. The inbreds were very weak plants* and produced rather small ears with few seeds. Thus, if inbred A were pollinated by inbred B, the seeds produced on A would be used for hybrids, but its small number of seeds made it of little practical use. Jones devised a double cross using four inbreds: A was crossed with B, and C with D, and the resulting hybrids AB and CD were then grown and a cross made between them. The resulting plant, being a hybrid, produced a large number of seeds, and these double-crossed seeds still produced plants showing extreme vigor and uniformity. By utilizing the double cross, it was now possible to convert a few bushels of single-crossed seed into one thousand bushels of double-crossed seed that could then be released to the farmer, making it economically feasible to grow hybrid corn. Thus, through the work of several scientists in addition to those mentioned here, and of practical farmers as well, hybrid corn became a reality. The double-cross method

---

* In fact, the inbreds were such sick-looking plants that in the early days of hybrid-corn development in at least one agricultural experiment station they were grown in out-of-the-way places where farmers would be unlikely to see them. The breeders felt that if the farmers saw the inbreds they would think the breeders were working in the wrong direction.

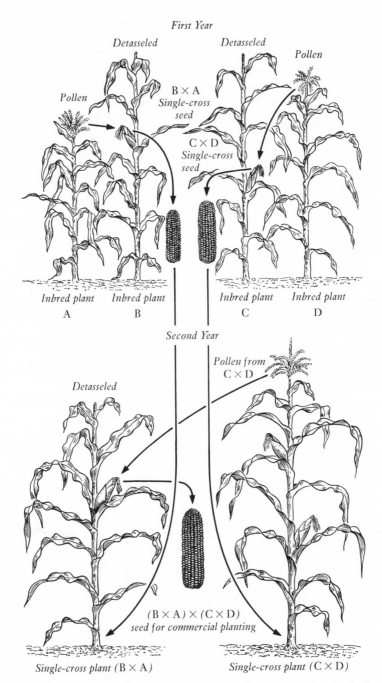

First Year

Detasseled *Detasseled*

Pollen *Pollen*

B × A
*Single-cross
seed*

C × D
*Single-cross
seed*

Inbred plant *Inbred plant* *Inbred plant* *Inbred plant*
A B C D

Second Year

Pollen from
C × D

Detasseled

(B × A) × (C × D)
seed for commercial planting

Single-cross plant (B × A) *Single-cross plant* (C × D)

Figure 5–20 Double-cross method of producing hybrid-corn seed. The inbreds shown at the top are crossed to produce seed that gives the single-cross hybrids shown below. The

for production of hybrid-corn seed was discontinued as better inbreds were developed. By 1970, 75 percent of the hybrid corn grown in the United States was derived from single crosses, and today nearly all of it comes from single crosses.

By careful selection of the inbreds, a breeder can produce high-yielding maize hybrids for many different climatic zones. Special characteristics, such as resistance to disease and tolerance to drought, can be incorporated, as well as special morphological features. Plants have been developed that produce two or three ears to the plant instead of one, uniformly placed on the stalk to make harvesting by mechanical pickers practical. Plants with stiff stalks have been produced to withstand mechanical picking. Hybrid-corn seed must be purchased anew each year, for if a farmer saved seeds from one year's crop for planting the next, he would lose much of the vigor and uniformity found in the original hybrids. Consequently, the production of hybrid-corn seed has become big business in the United States. Now, for the first time, farmers have to depend on outside sources for seed, which causes socioeconomic change in their lives.

The changeover to the use of hybrid corn was slow initially. Only 1 percent of the corn produced in the United States in 1935 was hybrid, but today virtually all the corn grown in this country is hybrid. Hybrid corn soon made possible the production on three acres what used to be produced on four. Equally important is the fact that hybrid corn has made possible a drastic reduction in the number of hours per person required to produce a bushel of corn. The increased yields with less human effort were particularly important during World War II, when, with a serious depletion of the labor force, the amount of corn produced decreased by only 10 percent. Since the war, yields have continued to increase dramatically, more than doubling in the space of twenty years. It should, of course, be obvious that other scientific advances have contributed to the great increase in corn production.

Hybrid maize has become important in other parts of the world and is gradually replacing the old types of maize grown in Latin America, with great increases in yield but also with great loss of diversity. Yugoslavia became a maize producer by importing hybrid-corn seed from Iowa. Seldom can a plant well adapted to one area be an immediate success in another, but Yugoslavia, being at the same latitude as Iowa and having a

---

single-cross plants are then crossed to produce double-crossed seed to be sold to the farmer. Single crosses have nearly entirely replaced double crosses in making hybrid corn today. (From "Hybrid Corn" by Paul C. Mangelsdorf. Copyright © 1951 by Scientific American, Inc. All rights reserved.)

climate not too dissimilar, was able to take immediate advantage of the introduced seeds.

In the production of seed for hybrid corn the seed companies grow a row of one inbred, A, to be the male parent, next to another, B, that is to serve as the female, or seed, parent. In order to insure that self-pollination of B does not occur, it is necessary to remove its tassels before pollen is shed. In the early years of seed production, this operation of detasseling was done by hand, students often being employed for the work. One of the recent advances in hybrid-seed production has been the elimination of detasseling through the discovery of plants that produce no pollen. The factor for pollen sterility, referred to as cytoplasmic male sterility, resides in the cytoplasm of the cell rather than in the nucleus and is inherited quite differently from ordinary Mendelian genes. Since plants derived from a male-sterile parent produce no pollen, it is necessary for a breeder to introduce a gene into the hybrids to make the next generation pollen-fertile. Virtually all of the hybrid seed sold in the United States by 1970 was produced by utilization of plants carrying the male-sterility factor along with genes for restoring male fertility when the plant was grown by the farmer. In 1970 the corn crop was decimated by a blight, which was found to occur only in those plants carrying the cytoplasmic factor.* For a number of years, therefore, the seed companies went back to the more laborious method of detasseling to produce hybrid seed, and some still do.

It will perhaps come as no surprise to learn that nowhere in the world is more maize produced than in the United States; Iowa and Illinois are by far the leading producing states. China is second in world production and Brazil is third. Maize is also of considerable importance in many other parts of the Old World, for in spite of the reluctance to accept it in some places, next to tobacco, it was the New World plant most widely adopted following the European discovery of America. Although its cultivation in the United States is concentrated in a relatively few states extending from western Ohio to eastern Kansas and Nebraska, maize is, by yield, the single most important crop plant in the United States. It ranks behind several other crops in export value, so obviously most of it is used at home. In fact, about half of the crop never leaves the farm but is used directly for livestock food. The small amount that does go to market is extremely important and yields more products than any other cereal. Practically all parts of the corn plant are used in some way, but the grain has the most significant use.

In the first edition of this book I reported that in a survey of a super-

---

*This also illustrates the great danger when a single variety comes to dominate a crop.

market it was found that corn was used in one way or another in the preparation of 197 different food products. Margaret Visser, in her recent book *Much Depends on Dinner: The Extraordinary History and Mythology, Allure and Obsessions, Perils and Taboos, of an Ordinary Meal*, goes beyond that, stating that one cannot buy anything at all in a North American supermarket that has been untouched by corn, with the occasional exception of fresh fish, and she goes on to justify her statement. Some two hundred million bushels go to the corn refineries, or "wetmillers," every year; here the oil-bearing germ, or embryo, is removed, then the outer covering, the gluten, and finally the starch, which constitutes more than half of the kernel. The starch is used directly, as cornstarch, and indirectly in food, drugs, cosmetics, and other products. Much of it is converted into syrups or sugars which find a great number of uses, for example, in foods, soft drinks, chewing gum, and in making cold-rubber tires. The annual consumption of corn sweeteners doubled from 1978 to 1986 to eighty-seven pounds per person, exceeding the amounts of beet and cane sugar consumed. The solubles from the steep water left after removal of other substances find many uses, as do the glutens—in animal feeds and in the manufacture of plastics, for example. The refined oil from the germ is used in cooking, in salad dressings, and in the making of margarine. Perhaps other plants, such as soybeans and coconuts, have more uses than maize, but certainly maize would run them a close race.

Research is constantly under way to find new uses for maize and to improve the plant, for maize has a badly balanced protein and, like rice, a lower protein content than wheat. It is also a poor source of certain vitamins. The deficiency disease pellagra, caused by a shortage of the vitamin niacin, is often found among people whose diet is inadequate. Large amounts of the amino acid tryptophan have a niacinlike effect on the body and may partially substitute for niacin. Since maize is deficient in both the vitamin and the amino acid, pellagra is common among some peoples who depend on it as their staple. How then did Native Americans fare so well with maize as their principal food? Throughout much of the Americas they prepared their maize using lime—from wood ashes, shells, or other sources. For example, tortillas are prepared by cooking the dried kernels in a weak solution of lime water for a half hour or more, after which the corn is ground to yield a *masa*, or dough. It has been shown that the use of heat with lime enhances the balance of the amino acids in maize and frees the small amount of niacin, which is otherwise unavailable. It is unlikely that Native Americans knew that this method of preparing maize increased its nutritive value; the procedure was probably adopted because they found that the hard kernels

Figure 5–21
Grinding maize, Mexico. (Reprinted with permission of The Macmillan Company from Paul Weatherwax, *Indian Corn in Old America.* Copyright © 1954 by The Macmillan Company.)

were softened by the lime water and could be ground more easily. How they happened on this happy discovery remains unknown. In 1963 scientists at Purdue University discovered that a mutant maize called opaque-2, which had been known for more than a quarter of a century, had a significantly higher lysine and tryptophan content than other types. The increase of these two amino acids vastly improves its nutritive value, so it was immediately recognized that the wide use of this new type of corn could help greatly to alleviate malnutrition in the Third World. Many problems were encountered, however. Yields were low and the plant was subject to fungal attacks. Moreover, the soft kernels made it unacceptable to many people and unsuitable for many industrial uses. Efforts to improve it persisted, particularly at CIMMYT* in Mexico, in spite of early discouragement. In 1988 it was announced that highly productive types, less subject to disease and having kernels practically indistinguishable from those of ordinary corn, had been created. Large-scale field trials are still needed, but it is predicted that "quality protein maize," as it is now called, could furnish ten million tons of quality protein to the world annually.

As a botanist, I cannot close this discussion without mention of another important, although sometimes overlooked, use of maize—in teaching and research. In the development of modern genetics, maize

---

*Centro Internacional de Mejormiento de Maíz y Trigo or International Maize and Wheat Improvement Center.

has been to the botanist what the fruit fly *Drosophila* has been to the zoologist. It is a particularly valuable plant because its chromosomes are exceptionally large, which has allowed detailed studies that have contributed greatly to our understanding of inheritance.

Sorghum

Grain sorghum or milo (*Sorghum bicolor*), a cereal somewhat resembling maize in vegetative features and bearing its grain in a terminal cluster, feeds millions of people in Africa and Asia. It is more drought tolerant than most cereals, hence it is of great importance in semiarid regions that will not support the growth of the major cereal plants. It is generally regarded as having been domesticated in Africa, where wild forms of the species are known; the savanna zone of eastern Africa, north of the equator, has been suggested as the place of origin. Some have held, however, that it was first domesticated in India. Archaeological sorghum grains from India date to 1800 B.C., far earlier than any archaeological material thus far reported from Africa. Only since 1950 has grain sorghum become an important crop in the United States, being grown mainly in the Southwest and used mostly for animal feed, and the country now leads the world in its production. Most grain sorghums

Figure 5–22   Grain sorghum. (Courtesy of FAO.)

Figure 5–23
Broomcorn. (Courtesy of FAO.)

grown here are hybrids developed through the use of cytoplasmic male-sterile lines.

In addition to the development of plants with large amounts of grain, *Sorghum bicolor* has been selected for other uses, in which the emphasis is on parts other than the grains. The sweet sorghums, or sorgos, which have a high concentration of sugar in the stem, are used for making syrups or for forage. Broomcorn is a sorghum grown for the stalks that are used to make brooms and brushes. The grass sorghums, such as Sudan grass and Tunis grass, have been selected for high yields of foliage, which is used for livestock feed. The grass sorghums are now placed in the species *Sorghum sudanense*.

Other Cereals

Three other important cereals are barley (*Hordeum vulgare*), rye (*Secale cereale*), and oats (*Avena sativa*). I have already mentioned barley as one

Figure 5–24 Field of barley in Holland. (Courtesy of FAO.)

of the first plants domesticated in the Near East.* In addition to being used directly as food for animals and plants, a large amount of barley goes to make malt for brewing and distilling. Malt is the germinated grain of barley. Rye and oats are both later domesticates and are thought to come from weeds that invaded the fields of the early farmers. As wheat and barley spread into areas where they were poorly adapted, the weeds thrived and eventually became intentionally cultivated. From the archaeological record it appears that rye was domesticated in Turkey and oats in Europe. As is well known, rye breads are common, but they usually contain high amounts of wheat flour, for rye has too little gluten to make a good bread by itself. Oats are one of the most nutritious of the cereals, the grain containing 15 to 16 percent protein and about 8 percent oils. They are particularly popular today as a breakfast food. At one time oats were a major crop in the United States for animal feed, but as tractors replaced horses on the farm in the first half of this century production fell drastically. Oats are still regarded, however, as one of the best foods for horses. Sales of oats for human food were brisk in 1988 when it was publicized that eating oat bran can reduce the cholesterol levels in the blood.

In addition to these grasses, there are a number of others, belonging to several genera and often collectively known as millets, that were domesticated in the Old World. Several of the millets are important locally in parts of Africa and Asia, particularly in areas where wheat and rice do not grow well. Job's tears or adlay (*Coix lacryma-jobi*) is still another cereal that was domesticated in Asia. Its grains are borne in hard beadlike structures formed from the leaf sheaths. These are sometimes strung and worn as beads. In the United States it is sometimes grown as an ornamental for its attractive "beads."

There are other plants, such as buckwheat, the amaranths, and the chenopods, with grainlike fruits that are used in ways similar to the cereals. They do not belong to the grass family and hence are often called pseudocereals. Buckwheat (*Fagopyrum esculentum*), an old domesticate of temperate eastern Asia, is perhaps best known in the United States as an ingredient in pancakes. Several species of both amaranth (*Amaranthus*), known as pigweed in the United States, and chenopod (*Chenopodium*), which includes our lamb's quarters and goosefoot, were

---

*In 1979 F. Wendorf reported barley from Egypt dated at 15,000 B.C., which would make it the earliest known domesticated plant. In 1984 (*Science* 225: 645) he published corrected dates on this material which are considerably later than the first reports of barley in the Near East.

Figure 5–25 Oats. The flower arrangement of oats readily sets it apart from the other small cereals, such as wheat, barley and rye. (Courtesy of USDA.)

domesticated in the Americas. Quinua (*Chenopodium quinoa*) is an old and important food source of the peoples of the high Andes and continues to be in wide use there. There has been increased interest in amaranths and quinua in recent years, for they both have been shown to have a good protein content with greater amounts of lysine than the true cereals. Both are now available in the United States, chiefly in health-food stores.

CHAPTER SIX

Sugar

Things sweet to taste prove in digestion sour.
WILLIAM SHAKESPEARE, King Richard II

Perhaps even more than meat, sugar is a luxury food. We can live without it, but it has always been in great demand because of the enjoyment it adds to eating. Two plants, sugar cane and sugar beet, furnish our table sugar. Many other plants also provide us with sugar. Two of these, maize and sorghum, were treated in the last chapter. Several palms provide sugar in southeastern Asia. Maple sugar from the sugar maple (*Acer saccharum*), indigenous in eastern North America, was an important source of sugar for Native Americans, and maple syrup is still much prized. Honey, perhaps the first known sweet, might also be considered a plant sugar in that bees manufacture it from the nectar of flowers. Several plants are known to contain substances that are several hundred or even more than a thousand times sweeter than sugar. One of these substances, thaumatin from the fruits of *Thaumatococcus daniellii*, native to tropical West Africa, is now available as a low-calorie sweetener in Japan and parts of Europe, but so far it has been approved only for use in chewing gum in the United States.

Sugar Cane

Although cotton is the plant most often associated with slavery in the United States, sugar cane was the plant most responsible for the establishment of slavery in the Americas. The need for labor to grow and harvest sugar cane in the West Indies led to the importation of large numbers of Africans at the beginning of the European colonization of the New World. Sugar cane very early became an important crop in tropical and subtropical America and today remains a major cash crop in many of the Latin countries, particularly Cuba and Brazil. In the United States it is grown in Hawaii as well as in some of the southern states.

Quite unlike the grasses (see Chapter 5), sugar cane (*Saccharum officinarum*) does not owe its importance to its seed and fruits. In fact this grass, which is valued for its stem, often sets few or no seeds. In prehis-

Figure 6–1 Sugar cane in flower, Costa Rica.

toric times people found that by chewing the stem they could obtain a sweet juice, as they still do in many places. Domestication is thought to have occurred in New Guinea or Indonesia, and the plant spread throughout much of southeastern Asia. Sugar cane did not become well known to the western world before the fifteenth century, and since that time it has spread throughout much of the world's tropics. Columbus is credited with having introduced it into the New World. Until the advent of the sugar-beet industry in the nineteenth century, sugar cane was the only source of sugar for most of the world's inhabitants, and today it still supplies over 50 percent of the world's sugar.

Sugar cane, a high polyploid species, is a tall plant, frequently reaching heights of ten to fifteen feet. Since it is a perennial, it will produce new stems for many years, but it is often replanted after the third harvest. Propagation is done vegetatively by using pieces of the stem. Extraction of the juice is accomplished by means of rollers. This is followed by purification that removes many of the "impurities," some of which are of nutritive value. The juice is then concentrated by evaporation. A boiling follows that leads to a crystallization yielding raw brown sugar and molasses. The brown sugar is then refined to produce white sugar. Except for the refining, all the steps in the processing of sugar are usually done on the plantation where the sugar is grown. The refining is frequently carried out in the importing countries of Europe or in the United States.

Refined sugar is one of the purest products to reach our tables and is one of our best energy sources. It has been said, however, that sugar contains "empty calories," which means that it supplies us with nothing but pure carbohydrate. A diet consisting mostly of sugar can lead to malnutrition, as in the case of the "sugar babies" of the West Indies, who are fed mostly on sugar and suffer from protein deficiency as a result.

Sugar cane is said to give the highest yield of calories per acre of any plant, and more and more of this energy is being used to propel automobiles rather than humans. Alcohol can be made from biomass, such as wheat or potatoes or even crop residues or garbage. In the United

Figure 6–2 Sugar cane for chewing, for sale in Mexico City.

Figure 6–3 Cutting sugar cane in Sri Lanka. (Courtesy of FAO.)

States some "surplus" grain, mostly corn, is being used to make ethanol, but some feel that to use grain for this purpose is immoral as long as there are hungry people in the world. Sugar cane, however, is not the basic staple that grain is. Moreover, overproduction has been a serious problem in many countries. Large amounts of sugar cane are used to make alcohol—*aguardiente* and rum—for human consumption, which some people, of course, also consider immoral. In 1975 Brazil launched an ambitious program to produce fuel alcohol from plants, particularly sugar cane, to reduce the need for expensive imported petroleum; in 1986 about one-half of Brazil's cars were run on ethanol. Ethanol as a fuel for automobiles is not new and in some ways is superior to gasoline for that purpose. It was first used in the last years of the nineteenth century in Germany and was made mostly from potatoes. For many years it has served as fuel in racing cars. As long as the price of gasoline remained low there was no incentive to develop ethanol for wider use as fuel. Even with the present cost of gasoline, it is still a much cheaper fuel than ethanol. In fact, in the United States more energy is required to make a gallon of ethanol than it yields in return. However, as the price of gasoline rises and methods of manufacturing alcohol from plants improve, the situation may change. Gasohol, a mixture of 10 percent ethanol and 90 percent gasoline, for example, became available to motorists in the United States in the 1970s.

Sugar Beet

The beet (*Beta vulgaris*) is an ancient food plant of Europe, probably first used for its leaves like the modern variety known as Swiss chard, but its use for sugar production goes back only two centuries. In the eighteenth century it was recognized that the root of the beet contained sugar, and attempts were made to increase its sugar content. A plant breeder, Achard, received help from the King of Prussia for sugar-beet produc-

Figure 6–4
Sugar beet.
(Courtesy of USDA.)

tion, and the first processing plant was built in Silesia in 1801. In the early part of the nineteenth century Napoleon, in spite of ridicule, encouraged the new sugar-beet industry in France. He realized that it could give his country a domestic source of sugar, thus freeing it from dependence on England, which had a monopoly on sugar cane. The sugar beet, a biennial or annual, has since become a major crop in the north temperate zone. France now leads the world in its production. Sugar-beet growing became established in the United States in the latter part of the last century and has become important in several states, notably Minnesota, California, Idaho, North Dakota, and Michigan. Sugar content has been increased from the 2 percent of the ordinary beet to around 20 percent. Tetraploid varieties have been created but have not proved as satisfactory as diploid ones; however, triploids have been more successful and are now widely grown in Europe. In spite of mechanization of nearly all of the operations connected with sugar-beet growing and harvesting, beet sugar has difficulty competing with cane sugar, whose production still employs much cheap hand labor. Price supports from the government for beet sugar are necessary in the United States. There is no difference in the sugars of the two plants; both are composed of over 99 percent sucrose.

The beet is a member of the Chenopodiaceae, or goosefoot, family. To the same family belongs spinach (*Spinacia oleracea*), of Old World origin, and quinua (*Chenopodium quinoa*), mentioned under the pseudocereals in Chapter 5. Another species of *Chenopodium*, known as Good King Henry, became domesticated as a pot herb in Europe, and various wild species of the genus, frequently called lamb's quarters, are collected in many places for the same use.

CHAPTER SEVEN

Legumes: The Meat of the Poor

And let them give us pulse to eat, and water to drink.

DANIEL 1:12

The first cultivated plants in the Old World, as we have seen, were grasses—wheat and barley—but they were soon joined by plants that added valuable supplements to the diet—lentils, peas, and vetches, all members of the botanical family Leguminosae. From the archaeological record we know that people were collecting seeds from wild plants of these species at the time they were beginning to cultivate the grasses and that their cultivation began not long after the domestication of the cereals. Cultivated lentils and peas appear in the archaeological record in the Near East nearly as early as wheat and barley. Evidence from the New World reveals that beans were among the first cultivated plants there, with two domesticated species appearing in Peru before 6000 B.C., antedating the appearance of maize. In the Far East the soybean, destined to become more important perhaps than any other legume, became an early domesticated plant.

It can hardly be an accident that these plants were among the first domesticates in so much of the world, for their seeds are an excellent food and among the highest of all plant foods in protein content. Even though the cereals are given the credit for making civilization possible, it must be pointed out that it could not have advanced nearly as rapidly without the legumes. Not only are the legumes high in protein, but their amino acids neatly complement those of the cereals, as was explained in Chapter 3; if we eat legumes and cereals together we obtain a far more complete protein than from eating any plant food alone. Thus wheat plus peas, maize plus beans, or rice plus lentils come close to filling our protein needs. But in a way, the domestication of complementary food plants must be partly a "happy accident," for primitive people knew nothing of proteins or amino acids, only that the seeds satisfied their hunger.

A passage from the King James Bible has been used to prove an early appreciation of the food value of the pulses, a name used for the edible-seeded legumes in England. After he was carried to Babylon, Daniel was ordered to eat the king's meat and wine. He resolved, however, not to defile himself by doing so and proposed a test. He and the servants would eat pulses and water while the other captive youths ate the king's meat and wine, and a comparison would be made at the end of ten days. At the end of that time, the Bible tells us, "their countenances appeared fairer and fatter than the children which did eat the portion of the king's meat." The argument for the value of the pulses loses its force, however, in modern translations of the Bible in which "meat" is given as "rich food" and "pulses" as "vegetables."

If we were to rank the families of plants in order of their importance to us, certainly the legumes would stand close to the grasses, for they serve not only as food but also, perhaps, in a greater variety of other ways than the grasses. One of the greatest services they perform is the fixation of atmospheric nitrogen, that is, the conversion of this element to a form available to other plants. When plants and animals die and decay, their nitrogenous compounds are broken down by bacterial action. Some of the nitrogen released is made available to other plants in the form of nitrates, but much of it escapes to the atmosphere and is not directly

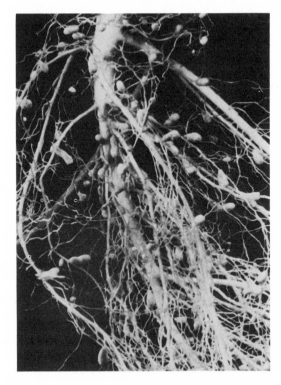

Figure 7–1
Legume root with nodules.
(Courtesy of USDA.)

available to plant life. Most members of the legume family, however, are able to obtain nitrogen from the atmosphere through the action of special bacteria that live in nodules on their roots. Very few other plants—no cultivated food plant among them—have this ability. Although the value of adding legumes to the soil had long been recognized, not until late in the nineteenth century was the scientific explanation for their importance discovered. The bacteria that live in the root nodules of legumes belong to several species of the genus *Rhizobium*; they derive their energy from their host plant and in turn provide nitrogen to the host in a symbiotic, or mutually beneficial, relationship. The fixed nitrogen goes into the production of protein, which is then available to humans through the seeds and to other animals through the leaves, stems, and seeds. When the plant dies, the nitrogen is returned to the soil, where it may be used by other plants. Although the nitrogen-fixing bacteria are widely distributed on wild and weedy legumes, farmers today, when planting a new field to legumes, often inoculate the seeds with a commercial preparation containing the bacteria to insure the presence of nodules or to increase their abundance. Before commercial fertilizer was widely available, legumes were frequently employed in crop rotation schemes to build up soil fertility and they are still so used in some parts of the world.

The legume family is an extremely large and cosmopolitan one, and all the major continents have provided members of it that have become food plants. These plants are most commonly known by the names pulses, peas, or beans, but not all plants called beans belong to this family; castor bean, for example, is not a legume. The seeds are the part most commonly eaten, and, like the cereal grains, most of them can be fairly readily stored for future use. Like the grasses, many of the domesticated legumes have lost their natural means of seed dispersal. Most members of the family produce a seed pod or fruit, called a legume, containing a row of seeds. In wild species the pod frequently is dehiscent at maturity, splitting along both sides, often forcefully expelling the seeds to some distance from the parent plant. Many of the domesticated forms have pods that are indehiscent*—an advantage to seed gatherers in that it makes it easy for them to collect all the seeds at the same time, but a disadvantage to the plant in that it interferes with the dispersal of seeds under natural conditions. The domesticated plants also have considerably larger seeds than their wild relatives.

---

*As a young professor of botany, I learned this the hard way. I brought green beans to class to illustrate the legume type of fruit, and I told my students that when they dried they would split open. They never did.

The food value of the seeds is high; they have about the same caloric value per unit weight as cereals and are a fair source of some vitamins and minerals. However, as already mentioned, it is their protein content that is striking, generally ranging from 17 to 25 percent, about double that of most cereals, to a high of 38 percent in the soybean. The protein quality is not as good as that of meat and other animal products, but meat is not regularly available to much of the world's population. The legumes are thus often thought of as poor peoples' food, and many of the world's poor could use more of them. Consumption, as might be expected, is highest in India, where both poverty and religious restrictions on meat contribute to their great use, and it is also very high in Latin America, where beans are commonly served with all meals. A disadvantage to using the legumes for food is that many members of this family possess antidigestive factors. Many beans contain undigestible material that induces flatulence.

In addition to the seeds, the green, unripe pods of legumes are often consumed, green beans being one of the most common vegetables in the United States. Less well known is the fact that the ripe pods of several of the tree legumes have a high sugar content and are eaten by humans and livestock. People today still enjoy the pods of carob, or St. John's Bread (*Ceratonia siliqua*), native to Syria, for their sweet taste, although they use them mostly as livestock feed; these pods are believed to be the "locusts" of John the Baptist in the Bible. The sugar obtained from the pods is used in preparation of chocolate substitutes, including a candy bar that, according to the advertisements, tastes like fine Dutch chocolate. The genus *Inga* in Latin America has extremely large pods, some nearly a yard long, containing seeds covered with a sweet white pulp. The pulp is sucked or chewed off and the seeds discarded. This was one of the few sweets known to the people of this region until the introduction of sugar cane. The pods are sold in many markets in tropical America, and the trees are among the preferred species cultivated to shade coffee plants. Young sprouts or seedlings of various kinds of beans are a popular food in some places, particularly in the Far East. Several legumes have edible roots; one, *Pachyrrhizus*, the yam bean or jicama, is cultivated primarily for its root in parts of Latin America as well as in the Orient, where it was introduced.

A legume that has recently drawn considerable attention is the winged bean (*Psophocarpus tetragonolobus*)—so called because its pod has winglike growths—an old domesticated plant of New Guinea. Nearly all parts of the plant can be eaten, and it has a protein content equal to that of the soy bean. Its cultivation has spread in much of the Old World tropics in the last ten years because of its high protein content. The soybean is not well adapted to tropical climates, and the winged bean may be an excellent substitute.

Legumes rival grasses as food for animals. When we speak of hay or forage, we generally are referring to both grasses and legumes but sometimes to other plants as well. Leaves of legumes are particularly valuable to animals for the same reason that the seeds are to people—their high protein content. Alfalfa, or lucerne (*Medicago sativa*), is one of the best, since it is both high yielding and high in protein and is probably the oldest cultivated forage plant. Seeds more than six thousand years old have been found in archaeological deposits in Iran. Alfalfa spread to Europe in early Christian times and in the last 150 years has made its way to most other parts of the world. Although it was introduced before the Revolutionary War, its importance as a crop in the United States dates from 1850, when seeds were brought to California from Chile by a gold miner. For a long time it was thought to be adapted only to the western states, but in this century varieties were found that would grow in other states, and now Wisconsin leads in its production. Alfalfa has been recommended as human food and is sold by some health-food stores, but it has never caught on—its flavor leaves something to be desired. Cigarettes made of alfalfa leaves have had the same fate for the same reason.

Other important forage plants of the legume family include several species of clover, as well as lespedeza, kudzu, vetches, and some of the plants also grown for food, such as the field pea. Most of these plants are also of value in erosion control and soil building, and they provide food for wild life, including nectar for bees. Although not equal to the grasses in controlling soil loss, they may serve as better soil builders because of their ability to fix nitrogen.

The family has also furnished us with some of our favorite ornamentals, such as sweet peas, lupines, and scarlet runner beans among the herbs, the vine wisteria, and a great many trees—some grown for their flowers, some for their graceful leaves, others for both. In the temperate zone of the United States the native red bud and honey locust and the Asian *Albizia* are extensively cultivated. The tropics furnish a great many more, many of them with brilliant flowers, such as poinciana. Some mention must also be made here of *Mimosa*, the sensitive plant, native to tropical America but widely grown in greenhouses for its sensitive leaves, which close to the touch. Countless students of botany have seen this plant used to illustrate plant movements.

A number of the trees, particularly tropical species, are valuable sources of wood, one of them being rosewood, much prized for making cabinets and other furniture. With firewood becoming scarce in the tropics, several other woody legumes, such as *Leucaena*, are becoming increasingly cultivated because of their rapid growth. Other members of the family produce gums and resins that are used in medicine and in varnishes. Dye plants are also provided by the legumes; indigo, until replaced by synthetics, was particularly valuable. Flavorings such as licorice and tonka beans are still of some importance.

On the negative side, in addition to providing some weeds, the legume family also includes a number of poisonous plants, some of which have caused deaths of both humans and domesticated animals. A few of these deserve our attention. The jequirity bean, or rosary pea (*Abrus precatorius*), a woody vine that is rather widespread in the tropics, has been newsworthy on several occasions in recent years, with warnings appearing about the toxic effects of the seeds. Beads made from the attractive bright red and black seeds have been brought home by tourists or imported for sale by stores. They make beautiful necklaces and are dangerous only if they are chewed and swallowed. The active principle, abrin, is one of the most toxic substances known, one seed containing enough to kill a person. It is reported, however, that in parts of Africa, the cooked seeds are eaten without harm! A few of the important leguminous food plants are known to be somewhat toxic under certain circumstances or to certain persons. The distinctive taste of the lima bean is due to a cyanide-containing compound (cyanogenetic glucoside), and, although beans grown and sold in the United States have only very small amounts of this substance, some varieties from the West Indies have enough to be considered dangerous.

Two diseases in humans, lathyrism and favism, are associated with legumes. The former results from consuming large amounts of the grass pea, or Indian pea (*Lathyrus sativus*). Eating the grass pea as part of an ordinary diet causes no harm, and the plant is in fairly wide use as food in both Asia and Europe. Difficulty usually arises when other food supplies are scarce and people are forced to subsist almost entirely on the legume. As the grass pea does well on poor soil and withstands drought, it may be in plentiful supply when other foods are lacking. Excessive consumption of it leads to a paralysis of the lower limbs that may be permanent. The disease has been most severe in India, but it was known to exist in ancient times in Europe. Favism, an acute anemic condition, results from eating the uncooked or only partially cooked broad bean (*Vicia faba*), also known as horse, Windsor, English, or fava bean, or from inhaling pollen from the plant. The disease affects only males of Mediterranean origin, and it is now thought that an inherited biochemical abnormality is responsible. Neither the grass pea nor the broad bean is eaten to any great extent in the United States.

The genus *Lupinus,* in addition to furnishing us ornamentals, has given us species used for food. Several species were domesticated in the Old World and one in the New for their seeds. Lupines contain alkaloids that are toxic, but inhabitants of the Old World selected alkaloid-free varieties for consumption by people and animals. One of these may be purchased at Italian markets in the United States under the name lupini bean. The Andean food species called *chocho,* or *tarwi* (*Lupinus mutabilis*), however, does contain large amounts of a bitter alkaloid. To make the seeds palatable as well as safe to eat, they are soaked in water for several days, which leaches out the alkaloid. The *chocho* was once an important protein source in the high Andes in the area where the starchy potato was the major food source. Although it is still fairly commonly cultivated in parts of the Andes, it has been replaced in many areas by the broad bean, which also does well at high elevations and does not need the lengthy preparation prior to eating.

Many of the wild plants in the family are also poisonous. Prominent among them are the locoweeds—various species of the genera *Astragalus* and *Oxytropis*—which are widespread in the western United States and have been known to cause death in livestock that have grazed on them. Although the family has supplied plants poisonous to humans, it has provided others that contain substances lethal to insects but relatively harmless to us. One of the safest insecticides, rotenone, comes from species of the South American genus *Lonchocarpus* and its Asiatic counterpart *Derris.* Long before their insecticidal properties were discovered, primitive people used these plants, as well as others, as fish poisons. The plants were pounded and placed in dammed waterways. The substance released from the plants stupified the fish, which were gathered as they came to the surface; the poisoned fish could be eaten without harm.

The list given here by no means exhausts all of the uses of members of the legume family, but it should serve to illustrate this plant family's broad significance. We may now return to some of the food plants and examine them in greater detail. Three of them, the common bean, peanut, and soybean, rank high among our most important domesticated species, and several others continue to occupy prominent roles as food plants.

As I have already pointed out, peas and lentils are among the very early cultivated plants of the Near East. They are still important there and throughout much of the world. The cultivated pea (*Pisum sativum*) comprises two main races or varieties—the field pea, now used mostly for forage and dried peas, and the garden pea with its high sugar content, considered by some to be the aristocratic food plant of this family.

Fresh peas were not popular before the seventeenth century, at which time they became esteemed in the court of Louis XIV. Whole pea pods used like green beans are a favorite food in the Far East and have now become popular in the United States. Biologists, of course, need no reminder that it was the pea that served as Mendel's research material for working out the laws of genetics.

The broad bean, previously discussed; chickpeas, or garbanzos (*Cicer arientinum*); and the cowpea, or black-eyed bean (*Vigna sinensis*), are other fairly important domesticates of Old World origin. The cowpea is widely used in the southern part of the United States. Another species, *Vigna sesquipedalis*, called asparagus, or yard-long, bean, is frequently offered by seed companies as a novelty, but those who grow it can hardly expect the beans to reach a yard in length. The specific epithet, translated as a foot and a half, comes closer to the truth.

Beans

No genus in the family has provided more edible species than has *Phaseolus*, and this is the group to which we generally refer when we speak of beans. Four different species were domesticated in the Americas. Two of these, the scarlet runner bean and the tepary bean, are of limited use today. The scarlet runner bean (*Phaseolus coccineus*), originally domesticated in Mexico, is grown in the United States mainly as an ornamental, but the large beans are a good food. The tepary bean (*Phaseolus acutifolius*), which was very important in prehistoric times in the American Southwest, is more drought resistant and less gas-forming than the other beans, but it is not grown commercially to any extent.

The lima, butter, or sieva bean (*Phaseolus lunatus*) is fairly extensively cultivated today. This species was first domesticated in South America, and a large-seeded form is known archaeologically from Peru, dated before 6000 B.C. The earliest remains from Mexico are dated much later, about A.D. 800, and are small-seeded forms. The plant that may be the ancestral type, while not common, is fairly widely distributed in tropical America today, and it is thought that the large- and small-seeded types of lima beans may represent independent domestications. Although the lima beans in markets in the United States are nearly always white when mature, a great variety of colors is known in those from Latin America. Speckled lima beans were used to convey messages in ancient Peru.

The most widespread and widely used bean today is *Phaseolus vulgaris*, known by a great variety of names—common bean, kidney bean, or, simply, bean being some of the most widely used in English. The snap, or green, beans and the wax beans are varieties of this species

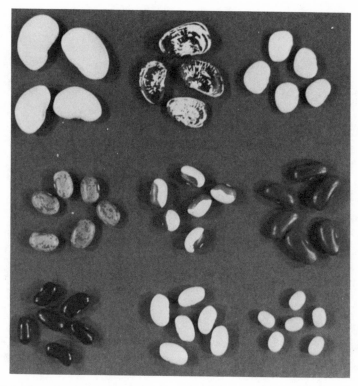

Figure 7–2 Some bean varieties sold in the United States. *Top row:* lima beans (*Phaseolus lunatus*)—Large Lima, Calico, Baby Lima. *Bottom two rows:* common beans (*Phaseolus vulgaris*)—*middle row,* Pinto, Yelloweye, Red Kidney; *bottom row,* Black Valentine, Great Northern, Small White.

whose pod is eaten in the immature stage. This bean was domesticated in Peru well before 6000 B.C. and in Mexico prior to 5000 B.C. Its wild ancestral form is widespread in tropical America, and it was independently domesticated in Peru and Mexico. From Mexico it spread northward, as did corn, and these two plants, along with squash, became the principal crops in North America in prehistoric times.

Today numerous varieties of the common bean are known; one estimate places the number at five hundred, only a few of which are grown in the United States. This bean is widely used in various parts of the world, and it still retains great importance in Latin America, where in different areas people have their own favorite varieties—red beans, black beans, or other types. Mashed beans for breakfast are not uncommon in parts of Latin America and beans, as previously mentioned, may be served along with rice at every meal. Because the plants are vinelike,

Figure 7–3 Harvesting green beans (*Phaseolus vulgaris*) in New York. (Courtesy of USDA.)

Native Americans commonly planted beans among their maize plants so that they could use the stalks to climb on. Many beans are still grown this way, but poles have replaced maize for supports in many areas. A mutant dwarf, or bush, type that needs no support has become a favorite in vegetable gardens in the United States.

The Old World beans formerly placed in the genus *Phaseolus* are now assigned to the genus *Vigna*. These include the black gram, or urd, bean; the golden gram, or mung bean; the adzuki bean; the rice bean; and the moth, or mat, bean. They are used mainly in India, China, Japan, and some of the neighboring areas. Their seeds are smaller than those of the American beans, and this may be one reason they have not been more widely adopted throughout the world. The germinated seeds of the mung bean are the widely used bean sprouts of Chinese cookery.

Peanuts

The two other legumes not yet discussed, the peanut and the soybean, are in many respects more significant than any of the foregoing. They have an exceptionally high protein content and a high content of oils, which have given them numerous uses in industry as well as being used for food. The peanut (*Arachis hypogaea*), also called groundnut, ground-pea, goober, pender, and many other names, is a rather unusual plant in that its fruits are produced under ground. The stalk of the flower elongates after fertilization and pushes the developing pod under the soil. The pod, or shell, is the fruit and the peanuts are the seeds.* For a long time it was thought that the peanut was indigenous to either China or Africa, but its true homeland is now clearly established to be South America. From there it was carried to other parts of the world in post-Columbian times. Its entry into the United States, however, was by way of Africa, for it was brought to Virginia by slaves. Archaeologically, the peanut is known from coastal Peru in the second millennium B.C., but recent botanical studies have indicated that the place of domestication was most likely the foothills of the Bolivian Andes. How early it was domesticated there and how it reached Peru pose interesting questions for the archaeologist and botanist.

Peanuts are reported in archaeological deposits in Mexico at Tehuacan in levels dated at about the beginning of the Christian era, but they never became as significant in Mexico as they did in parts of South America. The ancient Mexican name, *tlalcacahuatl*, means "ground cacao," presumably from the fruit being borne underground and its resemblance to the fruit of cacao, the chocolate plant. From this we get the modern Mexican name, *cacahuate*, although the word *mani*, which the Spanish picked up in the West Indies, is more widely used in other parts of Latin America.

Today the peanut, like its relative the soybean and many other crops—coffee, rubber, and quinine, for example—is grown more extensively in other parts of the world than in its place of origin. India and China lead the world in production of peanuts, followed by the United States and several African countries.† An attempt to grow peanuts on a huge scale in Africa led to one of the greatest agricultural fiascos in modern times. In 1946 the British government proposed to plant over three million acres of peanuts in East Africa for oil production. But clearing the land, irregular rainfall, and disease proved to be serious obstacles, and by 1951 only sixty-five thousand acres were under cultivation. Later the project was abandoned, but only after over a hundred

---

*Although they may be sold with the nuts at the grocery store, peanuts are not nuts, botanically speaking.

†A native African legume that also bears its fruit underground, the Bambara groundnut, was largely replaced by the peanut.

Figure 7–4 Peanuts on freshly dug plant. (Courtesy of USDA.)

million dollars had been spent. As John Purseglove has pointed out, this should be a lesson to those who would try to make abrupt changes in agricultural practice without sufficient research.

The peanut is put to far more uses in the United States than in any other part of the world. The plant is adapted to warm climates with light soil and is grown mainly in the southeastern part of the United States. It spread at a time when another crop was needed to replace cotton, which had been wiped out by the boll weevil. Only since 1850 has it become an important crop. In the last century most of the peanuts were used for roasting in the shell, and their prominence at circuses and baseball games can still be remembered. In fact, they were so much in demand that when one owner of a baseball park threatened to eliminate peanuts because of the work involved in cleaning up the shells, he almost had a rebellion on his hands. In the 1900s the mechanization of all operations connected with the growing and harvesting of peanuts was a strong impetus to increased production, and since that time peanuts have been a major crop in the United States. Of the numerous varieties of peanut only two, the Spanish and the Virginia, are commonly known in the United States. Georgia produces three times as many peanuts as any other state.

The principal use of the peanut is for human food, and a protein content of 26 percent makes it an extremely valuable one. In the United States the greatest amount goes into peanut butter, and countless children, of course, have subsisted almost entirely on peanut butter sandwiches, often by choice. Although indigenous South Americans had long ago made a similar product, its use in the United States dates from 1890, when a physician in St. Louis came up with it in a search for a nutritious, easily digested food for invalids, and from 1893, when Dr. John Harvey Kellogg, the health-food faddist of breakfast-cereal fame, made peanut butter so that some of his patients with poor teeth could take advantage of what he called "the noble nut." Peanut butter is very simply made; the best grades are made by grinding the roasted and blanched nuts and adding 1 or 2 percent salt. Stabilizers are sometimes used so that the oil does not separate out. Considerable amounts of peanuts, some seventy thousand tons annually, are also used in the manufacture of candy. Peanuts yield an oil that is esteemed as a cooking and salad oil and for the manufacture of margarine. There is considerable demand for peanuts in Europe, and in many years the United States produces a surplus that is used for export.

George Washington Carver was instrumental in developing new uses for the peanut early in this century, realizing that the welfare of the South could be greatly influenced by this plant. Today peanuts are important industrially and are used in the manufacture of a number of products, from shaving creams to plastics. The residue from the oil-extraction process can be used for fertilizers and livestock food and could, with proper treatment, be used for human food. Peanuts are an excellent food for fattening hogs, and sometimes the animals are turned loose in the fields to root out the nuts for themselves. The shells can be used to make insulating filler and wallboard, as well as in other ways.

Soybeans

Although the peanut has come to be one of the world's important food plants, its rise to prominence hardly equals that of soybeans. The soybean (*Glycine max*) can hardly be considered a new crop. Although it

A

B

Figure 7–5 *A.* Field of soybeans in the United States. (Courtesy of American Soybean Association.) *B.* Soybeans ready for harvesting.

C

D

C. Seeds of soybean. (Courtesy of American Soybean Association.) *D.* Harvesting soybeans. (Courtesy of American Soybean Association.)

didn't reach Europe before the eighteenth century and the United States until a century later, it was a very ancient and important cultivated plant in the Orient. Thus far there are no archaeological records to help us to establish when it was first cultivated, but its mention in Chinese literature before 1000 B.C. gives it a venerable age. The soybean is one of the richest foods known, containing 38 percent protein and 18 percent fats and oils. Unlike most members of the legume family, the beans are seldom eaten directly but are used for food, it is said, in four hundred different ways in the Orient.

The bean sprouts are used as a vegetable, and paste, curd, and "milk" made from the beans are used in a great variety of ways. For example, tempeh and miso, made by fermentation using fungi, and tofu, made by curdling the milk, have become fairly well known in the United States. Sauces of many different sorts are also prepared by fermenting the beans and adding other ingredients such as salt and wheat; sometimes even decomposed chicken or fish is added, which further enriches the protein content. Soy sauce has long been available in the United States, but it is prepared in a different way from that of the Orient and has a different flavor. In the Orient the milk has served in many of the same ways as cow's milk, for many people there are lactose intolerant. A popular soft drink, Vitasoy, is made from soybeans in Hong Kong. Today, however, China ranks only third in the world's production of soybeans. The United States has been the leading producer for a number of years, and Brazil, where the crop is grown in the temperate areas, is now a distant second.

The widespread cultivation and utilization of soybeans in the United States in the space of a few decades must be one of the most spectacular success stories in the recent history of agriculture. Soybeans were first grown in the United States in Pennsylvania in 1804, and although at the time it was mentioned that they should be more extensively cultivated, very little was done until the 1920s. Five million bushels were produced in the United States in 1925. Production reached 13 million bushels in 1933, 90 million in 1939, 299 million in 1950, 699 million in 1963, 1080 million in 1968, and over 2000 million in 1986. Little wonder that it has been called the "Cinderella crop." A number of factors contributed to the crop's rapid growth. The introduction of more than one thousand strains from the Orient in the late twenties provided material for the selection of improved varieties for different parts of the country. It was found that the soybean was well adapted to the corn belt, where it now ranks second only to maize.

Limits on the acreage of other crops by the government in the thirties encouraged the cultivation of the soybean. The development of specialized equipment for planting and harvesting was a factor in its acreage being increased, and the discovery that soybean meal was an excellent food for livestock, particularly poultry and hogs, made a large market readily available. At the same time, new food and industrial uses for the soybean came into being. Today the United States produces more than 60 percent of the world's soybeans, and about two-fifths of the crop is exported in one form or another, which makes it the United States' most important agricultural commodity in world trade.

Although it still has its greatest food use in the Orient, the plant has become increasingly used as food in other parts of the world, particularly as an additive for protein enrichment and as a meat substitute. By spinning the protein into long slender fibers and adding appropriate coloring, flavoring, and nutrients, soybeans can be made into synthetic beef, chicken, bacon, or ham. The taste, except possibly for the synthetic bacon, is not yet very close to the real thing, but improvements may be expected. The soybean is the world's greatest source of vegetable oil and the oil is of high quality. Most of it is used for the production of margarine, some for shortening and salad and cooking oil. Soybean lecithin, a substance yielded in processing the oil, is used in many ways—to preserve flavors of other foods, to help disperse nonsoluble compounds in food, and in whipped toppings, cake mixes, and instant beverages. Eighty-five percent of the soybean crop in the United States goes into the production of food for humans and livestock, and the remainder finds many uses in industry, perhaps more than any other plant. It is used to make a glue that is the most widely used adhesive in binding plywood, and in the manufacture of enamels, linoleum, printing ink, newsprint, and soaps—to name only a few.

Certainly many of the hungry nations could use this highly nutritious food, but at present they get little or none of it. Although the soybean has been introduced into many tropical areas in this century, it has as yet met with little success in most of them. In some Third World countries where it is now grown, the beans are used mostly for export, so it helps little in alleviating the local protein deficiencies.

CHAPTER EIGHT

The Starchy Staples

But don't forget the potatoes.

JOHN TYLER PETTEE, Prayer and Potatoes

Cereals, as we have seen, are the basic food. However, in former times and even today, those who live in areas poorly adapted to growing cereals have had to adopt some other plant as the mainstay of their diet. Several of these plants—chiefly the potato, the sweet potato, the yam, manioc, and the banana—are still extremely important.

Although not closely related, all of them belonging to different botanical families, these plants have much in common. All are tropical in origin, although the potato comes from the highlands, whereas the others are lowland plants. They are all propagated vegetatively rather than by seed. The archaeological record for most of them is scanty or nonexistent, as is true for most tropical plants, which grow in areas where conditions for preservation of plant remains are poor. All of them, with the exception of the banana, produce their edible parts underground. They provide extremely high yields of carbohydrates and thus supply much energy and a full belly, but all are sadly deficient in protein. A diet made up almost exclusively of any of them can lead to serious problems of malnutrition.

Potatoes

The white potato (*Solanum tuberosum*), also known as the Irish potato, rivals maize in volume produced and in value. It had a long way to go before it became an acceptable food plant in Europe, but few plants have figured more prominently in Western history than has the potato. The story begins in South America. Wild potatoes are fairly widespread in the Americas, particularly in the Andes. They were probably found to be a valuable food when people first entered this area, and at some undetermined time, probably more than four thousand years ago, they came to be intentionally cultivated. With intentional selection there was an increase in size, and the potato became the most important food plant in the high Andes, for it thrives at an elevation where few other cultivated

Figure 8–1 White, or Irish, potato plant. (From "The Late Blight of Potatoes" by John E. Niederhauser and William C. Cobb. Copyright © 1959 by Scientific American, Inc. All rights reserved.)

plants will grow. Maize will not grow at elevations much higher than eleven thousand feet and is not a particularly productive plant at that altitude, but potatoes do well at fifteen thousand feet. Frost, common in parts of the Andes, is not conducive to keeping potatoes, but the native South Americans found a way to make it an ally. They allowed the poatoes to freeze at night and then stamped on them as they thawed the next day. This process, repeated for several days, removes the water and results in a desiccated potato, called *chuño*, which may be kept almost indefinitely and used as required. There was another reason, perhaps the original one, for preparing potatoes in this fashion. The wild potatoes of the Andes contain alkaloids which make them toxic and quite bitter to the taste. By freezing and stamping the potatoes, the alkaloids are largely eliminated. Thus was born one of the original "freeze-dried" foods, in a sense the forerunner of instant mashed potatoes but with a somewhat different taste. Although the flavor of *chuño* is not pleasing to all foreign visitors in the Andes, it is still the staff of life to many people in highland Peru and Bolivia. The potato was cultivated throughout the length of the Andes in prehistoric times, but it did not make its way to Central America until it was introduced there by the Spanish.

The first European on record to see the potato in America thought that it was a strange food and compared it to the truffle. Potatoes were introduced into Europe in 1570, but their acceptance was far from immediate. The Jerusalem artichoke, introduced from North America at about the same time as the potato, was soon regarded as food fit for a queen; today it is seldom used for food, while the potato prevails. The slow acceptance of the potato was due to several factors. The newly introduced plant probably was not very productive at first, and the fact that it was recognized as a member of the nightshade family may have contributed to a reluctance to accept it. At this time the nightshade family was known in Europe mainly through its poisonous members— mandrake, henbane, and belladonna—and the family had not yet provided any important foods. Even after the potato gained some acceptance as food, there were many who disapproved of it. One minister preached against it, stating that if God had intended it to be eaten it would have been mentioned in the Bible. Several writers of the time condemned it for its flatulent or "windie" property, among other reasons. At the start of the sixteenth century it was thought to be an aphrodisiac, a notion stemming from its confusion with the sweet potato, but this, of course, may have increased its use among some people. Promoted by royalty in some countries, the potato gradually became more widely grown. Various wars that destroyed standing crops of other food plants helped to increase its popularity, for the potato tubers, safe underground, could not be destroyed by burning as easily as a field of grain. The potato did not become common in Europe before the eighteenth century, and there can be little doubt that it contributed to the increase of the population at that time, not because of any aphrodisiac property but because of its food value. World War I has even been blamed on the potato, since this food was responsible for the increased population that has been considered a contributing factor to the war. Certainly it was the potato that helped to keep Germany alive during two world wars.

In the early part of the nineteenth century the potato had become the dominant food in Ireland. In fact, it was almost the sole food of the peasantry; the average consumption of potatoes was 10 to 12 pounds per person a day. When a blight struck in 1845-46, wiping out nearly the entire crop, famine followed. An estimated one and a half million people died as a result, and another million emigrated, many of them to the Americas. The blight disease was not understood at the time. Only later was it recognized that the cause was a fungus. Since that time plant pathologists and breeders have devoted a great deal of effort to controlling the disease and developing resistant varieties of potatoes. Much has been accomplished, but blight still causes serious losses in many parts of the world. It was formerly thought that eating blighted potatoes had no ill effect on humans, but recently it has been shown that there is a high incidence of certain kinds of birth defects in areas where potato blight is common. In an experiment, marmosets that ate blighted potatoes as part of their ration produced some offspring showing birth defects; a control group of the animals, fed an ordinary diet, bore all healthy young.

Although the potato is often referred to as a root crop, the part eaten is actually an underground stem, specialized for food storage and known botanically as a tuber. The "eyes" of the potatoes are buds, and the so-called "seed potatoes" used for vegetative propagation are not seeds at all but portions of tubers, each containing an eye. Potato plants rarely flower in some northern areas, but they are capable of flowering and do so throughout much of their range. The attractive white, blue, or pink flowers produce small green berries, something like a small, unripe tomato, that contain seeds. Although these true seeds are now available for growing potatoes, most of the world's potatoes are still grown vegetatively from pieces of the tuber. Seeds, however, are very important to plant breeders in creating new varieties. Vegetative reproduction, which is asexual, gives rise with rare exceptions to types exactly like the parent. Sexual reproduction by seed, however, allows for the production of new genetic combinations, some of which may be superior to the parents.

The potato tuber, like most vegetables, is mostly water; it contains 17 to 34 percent carbohydrate, small amounts of protein, a trace of fat, and some vitamin C. Varieties commonly cultivated in the north temperate zone contain from 1 to 3 percent protein, but some varieties grown in the Andes have 6 or 7 percent. Although potatoes contain less protein than cereals, they yield more than twice as many calories per acre than wheat and rice. Many of the nutrients in the potato lie next to the skin; these are removed and therefore lost through the practice of deep peeling. The potato reveals its relationship to the other nightshades not only by its flower and fruit but also in that it may contain the toxic alkaloid solanine. Potato tubers grown exposed to light turn green and produce this alkaloid. Eating of the leaves or other green parts of the plants has caused posioning in livestock and humans. In fact, at times even the eating of the tubers has caused severe gastric distress. Lenape, a variety released by the U.S. Department of Agriculture in the sixties, was later found to contain more than twice the normal amount of alkaloids; in 1970 it was withdrawn from cultivation.

The English name *potato* for this plant is a mistake that goes back to its introduction into England. The word potato is actually derived from *batata*, a Native American name for the sweet potato. There was considerable confusion over the various new root crops being introduced into Europe in the sixteenth century, and the name of the sweet potato became attached to the white potato, where it remains.

Today over three hundred million tons of potatoes are produced annually. The Soviet Union is the largest grower, providing nearly 30 percent of that amount; China is second and Poland is third. The United States is far behind with only 5 percent of the world's production. In addition to their use as food for humans, large amounts of potatoes are fed to livestock in Europe and some go into the production of starch, used chiefly for sizing cloth and paper. Potatoes also serve as a source of alcohol, both for drinking and industrial purposes. The potato is generally thought to have reached the United States by way of Bermuda in 1621, although its start as a crop apparently stems from an introduction to New Hampshire from Ireland in 1719. In the United States nearly one-third of the harvest is grown for direct human consumption, and about half of the crop reaches the table in other ways as potato chips, frozen french fries, and dehydrated potatoes. Small amounts are used for livestock and starch; the bulk of the starch produced in the United States, however, is made from corn. Idaho is the leading state for the production of potatoes. The potato, of course, is still widely grown in the homeland, the Andes, and in Mexico and Central America, where it was introduced by the Spanish. It has been grown in India, which now produces as much as the United States, since the seventeenth century, and it later reached Africa, where it is now widely grown particularly in the highlands. It has not produced well in the lowland tropics, where its role as human food is taken over by some of the other starch crops, particularly manioc. The International Potato Center, located near Lima, Peru, has for several years worked at producing potatoes that will yield well in the lowland tropics. The center has over five thousand kinds of potatoes in its collection, a valuable resource for the potato breeder.

Along with the effort to improve the potato, considerable study has also been devoted to its classification, origin and evolution. *Solanum tuberosum* comprises two main races or subspecies. The potato that was originally introduced to Europe and spread around the world was from the northern Andes (*Solanum tuberosum* subspecies *andigenum*). The blight largely eliminated this potato not only from Ireland but from other areas as well. According to Paul Grun, potatoes from Chile (*Solanum tuberosum* subspecies *tuberosum*) were introduced into Europe and North America in the late part of the nineteenth century, and it is these and hybrids of them with *Solanum tuberosum* that constitute most of the potatoes today. *Solanum tuberosum* is a tetraploid species, and, while there is still not complete agreement, considerable progress has been made in determining what diploid species give rise to it.

In the spring of 1970 five million pounds of potatoes were destroyed by fire in eastern Idaho. It was no accident, for the fire had been deliberately set by farmers in an attempt to raise the price. The same news report also carried the information that an estimated fifteen million Americans are underfed.

Sweet Potatoes

At the time of the European discovery of the New World, the sweet potato (*Ipomoea batatas*) was widely cultivated in tropical America and was also being grown on some of the Pacific islands. In fact, the sweet potato had become the principal food of the Maoris in New Zealand. The presence of the plant on either side of the Pacific at such an early date poses several interesting questions—among them, how and when did it get across the ocean? The sweet potato is propagated vegetatively from stem cuttings or root sprouts. The plant under proper conditions may flower and set seeds, although, as with the white potato, seeds today are used only in breeding. It has been pointed out, however, that seeds, carried either by birds or on floating logs, may have reached the Pacific islands and given rise to plants that were later discovered and cultivated. The suggestion has also been made that the sweet potato might have been independently domesticated in the Americas and in

Figure 8–2 Sweet potatoes on vine.

the Pacific area from similar wild species. While either the introduction of seeds by some natural means or an independent domestication remains a possibility, it seems far more likely that people were responsible for the introduction of the sweet potato from the Americas to the Pacific region. There are two ways in which this might have occurred.

Ancient Polynesian voyagers, who were efficient sailors, could have traveled to South America and picked up sweet potatoes, then carried them home and cultivated them. Or, and perhaps more likely, Peruvians, who also are known to have had seaworthy vessels, may have carried the sweet potato to one of the Polynesian islands, not necessarily intentionally but perhaps when driven off their normal course by winds. The prevailing currents are in the right direction for such a trip, and there is no need in this case to account for a return voyage. There may be some question about whether the sweet potatoes would not have been eaten on the journey and, if not, whether they still would have been viable after such a long voyage, for sweet potatoes are rather perishable.

The presence of the sweet potato in both the Pacific region and South America does not necessarily mean that voyages across the Pacific were frequent in prehistoric times. If they had been, we would expect other plants to have been exchanged among the peoples of the two areas.

Many such claims have been made for the spread of other plants, but except for two of them the evidence is unconvincing. The only plants besides the sweet potato that did become established on both sides of the Pacific in prehistoric times are the coconut and the bottle gourd, both of which have fruits that float and hence are well adapted to dispersal by ocean currents. The voyage of the *Kon Tiki* by Thor Heyerdahl was an attempt to prove that primitive people could have crossed the Pacific successfully, and Heyerdahl and others have claimed that there were significant contacts in prehistoric times with an exchange of cultivated plants. There may have been such voyages, but as far as plants are concerned, the only support for the idea of exchange appears to be provided by the sweet potato, and a single accidental voyage could account for that.

Although the origin of the sweet potato has recently received considerable attention, much of its early history in the Americas remains obscure. The sweet potato, like bread wheat, is a hexaploid. An allegedly wild species, *Ipomoea trifida*, also a hexaploid, has recently been reported from Mexico and is considered by Ichizo Nishiyama to be the ancestor of the domesticated plant. If this is true, it would seem likely that the sweet potato had its origin in Mexico. However, some botanists feel that *Ipomoea trifida* is an escaped form of the domesticated sweet potato and not a truly wild plant. The lack of popularity of the sweet potato in Mexico suggests that it may have originated elsewhere, and anthropological and linguistic considerations indicate that this may have been in northwestern South America. In this connection it is of interest that archaeological remains of the sweet potato are known from Peru but have yet to be found in other areas. The plant was being grown in the Caribbean area as well as in mainland America when the Spanish arrived. The exact place or places of its origin as well as the history of its dispersal remain to be clarified.

The post-Columbian history of the sweet potato, as is to be expected, is much better known. Columbus took the plant to Spain, and it later became a common staple on ships on return voyages to that country. The Spanish also introduced it from Mexico to Guam so that it might be available for ships' supplies on voyages to the Far East. As the plant is adapted to warm climates, it was not cultivated in most of Europe. Sweet potatoes grown in Spain were imported into England, where for reasons not entirely clear they were regarded as an aphrodisiac. Humankind has always wanted to attach aphrodisiacal properties to plants, and it is only natural that such properties would become associated with exotic plants instead of those already well known.

The plant had various names in the Americas—*apichu* among the Quechua of Peru, *camote* in Mexico, and *aje* for starchy types and *batata* for sweet types in the Caribbean area.* Of these, the last became the most generally accepted and was transferred to the white potato as related above, so that we have to refer to the plant under discussion as the sweet potato. This name, of course, is not inappropriate, since 3 to 6 percent of the carbohydrate is in the form of sugar, and the amount may increase in storage and with cooking. The plant provides 50 percent more calories than the white potato but generally has less protein (1½ to 2 percent). It is a good source of vitamin A and of minerals. In contrast to the white potato, the part eaten is a true root.

Although the sweet potato is still widely grown in the Americas and has become an important crop in the southern United States, it is now more extensively cultivated in Africa and Asia; China leads the world in its production. Although its importance in Japan has declined in recent years, that country still produces three times as much as the United States. In addition to its use for human and livestock food, considerable amounts are used for alcohol production in Japan. In both Japan and Taiwan, the crop is regarded as "typhoon insurance," for when rice and other standing crops are destroyed the sweet potatoes will still be available for food.

The sweet potato is a member of the morning-glory family. The flowers, seldom if ever produced under temperate-zone conditions, bear a close resemblance to the ornamental forms of the morning glory. Although more common at lower elevations, the plant can be successfully cultivated as high as nine thousand feet in the tropics. The roots cannot stand waterlogging, and for this reason the plants are often grown on ridges or mounds to provide good drainage. The vines rapidly cover the ground and hence the plants require little cultivation. In spite of the fact that for agricultural production the crop requires good drainage, the roots will sprout readily if placed in a jar of water, and the vines will grow luxuriantly; not infrequently such plants are grown as ornamentals in the home. In some areas the leaves as well as the roots have served as food.

---

*The name *kumar*, or variants of it, was used both in Peru and Polynesia for the sweet potato. Until recently it was claimed that this was further evidence that the sweet potato was introduced from Peru to the Pacific area, for it would be most unlikely for the same name to have been independently chosen in the two areas. The Peruvian origin of the name has been questioned in recent years, and it has been postulated that *kumar* was actually of Polynesian origin and was introduced into Peru in the early post-Columbian period.

Figure 8–3   Manioc plantation, Ecuador. (Courtesy of FAO.)

Manioc

Manioc, cassava, and yuca* are some of the common names of *Manihot esculenta*. This plant is little known to most people in the temperate zones except in the form of tapioca, although it is one of the extremely important food plants of the tropics of both hemispheres. Although manioc is New World in origin, details about where and when it was first domesticated remain vague. Apparently wild forms of manioc are known from South America, and archaeological material has been recovered from sites in Peru. Both Venezuela and Brazil have been mentioned as places of origin of the domesticate. The plant may have been independently domesticated in Middle America, but an introduction from South America to Middle America seems much more likely. Manioc belongs to the Euphorbiaceae, or spurge, family, which includes, among many other plants, the Pará rubber tree, our best source of natural rubber, and the poinsettia, a well-known Christmas ornamental. The manioc plant is rather tall, at times reaching fifteen feet, with divided leaves. The edible part is the tuberous root, which somewhat resembles a sweet potato but is usually much larger—some grow to be a yard long and weigh several pounds. Numerous rather ill-defined varieties exist that are generally divided into two groups, the "sweet" maniocs and the "bitter." The latter contain higher concentrations of poisonous cyanide compounds (cyanogenetic glucosides) than the for-

*Not to be confused with *Yucca*, an entirely different plant.

Figure 8–4 Manioc roots (bottom left) for sale in Ecuadorian market. Also shown are fresh maize for roasting on the cob, limes, and small hot chili peppers (bottom center).

mer and require special preparation by grating, pressure, and heat to make them safe to eat. One wonders, of course, how anyone discovered that this plant, as well as several others that are toxic until specially prepared, could be made edible.

Manioc was taken to Africa from Brazil by the Portuguese in the sixteenth century, but it did not spread widely there until the twentieth century, when its cultivation was encouraged. It was found that manioc was not damaged by locusts, a serious pest of crops in many parts of Africa, and that the bitter varieties could be grown in areas where wild animals would destroy other crops. The plant also grows better on poor soil than any other major food plant. As a result, Africa today leads the world in its production, which in one sense is unfortunate, since the roots contain little protein and their wide use has contributed to malnutrition. The plant reached the eastern tropics somewhat later than it reached Africa, and presently is most important in Indonesia, where it ranks as the second most important crop after rice. Some is grown there for export, but most of it is used locally for food, as is true in most other areas where the plant is grown.

Manioc is a lowland tropical crop, although sometimes it is grown at elevations as high as six thousand feet. It will grow in somewhat arid regions as well as in regions with fairly high rainfall. Stem cuttings are used for propagation; these are simply stuck in the ground at an angle in fields prepared by slashing and burning. The plants are then more or less left to themselves. In some varieties the roots mature in as little as six or seven months, and in others roots may continue to increase in size for up to four years. The roots are harvested as needed at the farmer's convenience. The plants are extremely productive.

The peeled roots of the sweet types of manioc may be prepared for eating simply by boiling or roasting. Both sweet and bitter varieties may be used to yield a coarse meal, known as *farinha de mandioca* in Brazil. The meal is often prepared by placing cut roots into a long sleevelike basket, known as *tipiti* in Brazil. The *tipiti*, which works something like a Chinese finger lock, is tied to a tree and pressure is exerted on the other end. The pressure extracts the juice, which is also collected and is often used to prepare sauces or beers. Among some peoples in the lowland regions of tropical South America, the beer is prepared by old women who sit around a large gourd vessel, chew the roots, then spit them into the gourd. The chewing initiates a breakdown of the starch into sugar, and wild yeasts then take over the production of alcohol. Visitors to a tribe are often expected to take a ritual drink of the beer, and to refuse to do so would be considered an insult.

At one time manioc was in demand in the United States in the form of tapioca. Once a popular pudding, tapioca has been largely replaced by gelatins and instant puddings. The fact that some people referred to tapioca pudding as "fish eyes," a fairly apt description, probably did not figure in its decline. Tapioca, as purchased for making desserts, is prepared by gentle heating, the partial cooking causing the agglutination of the manioc starch into small pellets. In addition to its use as food, manioc starch is used in the manufacture of adhesives and cosmetics, for sizing textiles, and in making paper. Since many of its former uses have now been taken over by starch from waxy maize, little manioc now enters into international trade.

In some places, particularly in Africa, manioc leaves are used as a pot herb. Since the leaves may contain up to 30 percent protein, their wider use might help prevent malnutrition among manioc-root eaters. Manioc improvement programs, which are now under way in both Africa and South America, are attempting to increase yield and disease resistance. Attempts to increase the protein content have thus far met with little success. Most varieties have 1 percent protein or even less in the roots, but a few have been reported to have nearly 3 percent, although these high protein types produce small, somewhat woody roots. It could well be that roots and tubers are designed by nature to store starch rather than protein, so efforts to increase protein content may be in vain.

Yams

In prehistoric times the most widely distributed of the starchy crops were the yams, various species of the genus *Dioscorea*. There is no need to call upon humans as an agent for their very wide dispersal, for the genus contains some six hundred species, native to the tropics of both hemispheres. People in many different areas discovered independently that the large underground stems, or tubers, were a good source of food. The tubers of some species under cultivation may reach a remarkable size, six to nine feet and weighing more than one hundred pounds.

The true yams are largely confined to the tropics and are little known in the United States. Most of the so-called yams in markets in the United States are moist-fleshed varieties of sweet potatoes. One true yam species, *Dioscorea bulbifera*, is sometimes cultivated in greenhouses under the name "aerial potato" or "yam potato." The small aerial tubers, or bulbils, produced on this vine are sometimes used as food in parts of Asia and Africa.

Yams grow best in humid and semihumid regions. They are usually grown on a small scale by farmers for their own use, frequently in shifting cultivation, so that new fields are sought after a few years. Tuber cuttings, small tubers, or bulbils are used for planting. The plants are usually grown in ridges or mounds, and stakes are often provided as supports for the vines. The harvest season, which is an important occasion to those people for whom this plant is the major food source, is celebrated with special rites. After harvest, the tubers are stored; they are later boiled, roasted, or fried as they are needed.

Today the greatest production is in West Africa, where in many places the yam is the principal food plant, as it is in parts of southeastern Asia, which is second in production. Large quantities of yams are also still cultivated in the Caribbean. The species introduced from the Old World, which first came to America as food supplies in slave ships, are now probably more extensively cultivated in the West Indies than is the native American plant.

Yams, however, are no longer as important in the Old World as they

Figure 8–5 A. Greenhouse plant of potato yam (*Dioscorea bulbifera*) showing the aerial "potatoes," or tubers. *B.* Tubers of *Dioscorea alata.* Some tubers of this yam grow to be several feet long. (Courtesy of USDA.)

A

B

once were, chiefly because of the introduction of other tuber crops, particularly manioc. On the one hand, this is unfortunate, for yams have a higher protein content than manioc. On the other, the large amount of manual labor required to grow yams makes the crop a relatively inefficient one in terms of food yield for the hours spent on it. Little work has been done in an attempt to improve the yam, chiefly because it is a crop that is consumed locally and does not enter into trade with the developed nations. The cultivated plants are little used except as a source of food.

Some of the wild yam species have also been used as food in periods of famine by some people in the tropics. Many of the wild species, however, contain toxic substances and require special treatment to make them safe to eat. They have been known to cause deaths in humans. It is some of these same toxic substances, however, that have made the wild species useful in some other ways. Some of them have been used as fish poisons, similar to the way rotenone is used, and around 1940 some steroids in *Dioscorea* were found useful in the manufacture of cortisone and sex hormones. At one time the steroids useful in the treatment of Addison's disease, asthma, arthritis, and skin disease were thought to occur only in animals, and their production was very expensive. As a result of the discovery of plant sources, the cost of hormones fell from $80 a gram to $2 in ten years. A still more significant use became known in 1956 when Dr. Gregory Pincus announced that a drug derived indirectly from *Dioscorea* would stop ovulation and hence prevent conception. Up to that time steroids that prevented conception had to be taken by injection, whereas it now became possible to use oral administration. Tests of the new drug in Puerto Rico and Los Angeles were successful, and the birth-control pill was on its way. Although most birth-control pills are wholly synthetic today, *Dioscorea* still figures in their origin, and in this way the plant has contributed more to controlling the world hunger problem that it will ever do as a food.

Taro

Another tropical crop that feeds millions of people is taro, or dasheen. It was probably originally domesticated in southeastern Asia, although our earliest historical record comes from China. Details about its origin have not yet been worked out, but it is known to have been carried quite early from its homeland to Japan. It was also introduced into the Pacific islands, apparently by Polynesians, and there it is still a staple in some areas. Taro eventually reached Africa, which today leads the world in its production, and was carried from there by slaves to tropical America. It was introduced into the southern United States in 1910 as a crop in soils too moist for potatoes, but it never made much impact, for it couldn't be grown economically enough to compete with other root crops.

Taro (*Colocasia esculenta*) is very similar in appearance to elephant's ear, a plant grown as an ornamental or a curiosity for its extremely large heart-shaped leaves. These plants belong to the Araceae, or aroid family, which perhaps is best known to most Americans through *Philodendron*, widely grown as a house plant. Members of this family usually contain crystals of calcium oxalate in nearly all parts of the plant, and these can be toxic. Anyone who has ever bitten into the tuber of Jack-in-the-pulpit, another member of this family, is familiar with the action of these crystals. The effect might be compared to biting into a pincushion with pins present. Fortunately, the calcium oxalate crystals are usually destroyed by boiling. The American counterpart of taro is yautia, that is, any of the various species of *Xanthosoma*, an ancient cultivated plant. Both it and taro are cultivated in parts of lowland tropical America today.

Leaves of taro are eaten, but the part usually consumed is the underground portion, known as a corm and made up mostly of stem tissue. The corms contain about 30 percent starch, 3 percent sugar, and a little more than 1 percent protein; they are fairly good sources of calcium and phosphorus. Reportedly, thousands of varieties are known, with the color of the corms ranging from white to yellow and pink. A pink-fleshed variety, which is one of the favorites today, reputedly was reserved for royalty in early times in Hawaii. One of the favorite uses there, then as now, was to make poi. Steamed corms are crushed, made into a dough, and allowed to ferment for a few days. The dough is then eaten by dipping into it with the fingers or rolling it into small balls. People used to eat ten to twenty pounds of poi a day. Commercial preparation has now largely replaced the making of poi at home. Not all visitors to the islands appreciate the taste of poi, some comparing the flavor and consistency to paste.

The Hawaiian luau gets its name from the leaves of taro, which are used as part of the meal. The leaves are a good source of vitamin A and C and undoubtedly contain considerably more protein than the corms. The leaf stalks are a favorite food in much of Polynesia. A taro flour, taro chips, and breakfast foods have been made from the corms in Hawaii. Since taro is easily digested, it has been recommended for use in baby foods. Today, however, there is little use of processed taro outside of Hawaii.

Taro is one of the few important cultivated plants that thrive in wet soil, although certain varieties can grow in relatively dry areas. The

A

B

Figure 8–6 *A.* Taro plant. (Courtesy of FAO.) *B.* Taro corms. (Courtesy of USDA.)

ancient Hawaiians accomplished some rather remarkable engineering feats to provide suitable areas for its cultivation. The propagation is vegetative, from corm tops or axillary corms, since the plant rarely flowers and seldom sets seed. The statement may sometimes be seen in older books that the plant has been in cultivation so long that it no longer flowers. Although the exact cause of the failure to flower is not known, age almost certainly has nothing to do with it. It more likely reflects a hybrid origin, mutations, or the fact that the plants are cultivated in areas where the day during the growing season is of an improper length to induce flowering.

Breadfruit

Another plant that has served as a staple, although not as important as any of those already discussed, is the breadfruit (*Artocarpus altilis*), a member of the Moraceae, or mulberry family. The breadfruit is a handsome tree, forty to sixty feet tall, with shiny, deeply lobed leaves. The large fruits in reality are multiple fruits, since each develops from the ovaries of a tight cluster of flowers rather than a single flower. Sometimes reaching a foot in diameter and ten pounds in weight, the fruits are a rich source of carbohydrates and have been used as a food in Polynesia since prehistoric times.

Captain James Cook saw the tree in his voyages in the Pacific, and from his descriptions some Englishmen thought that breadfruit would make a wonderful food for slaves in the West Indies. Captain William Bligh, who had sailed with Cook, was commissioned to bring trees to the West Indies from Tahiti. Thus began the famous voyage of the H.M.S. *Bounty* in 1789. The ship left Tahiti with more than a thousand young trees, but as a result of the mutiny led by Fletcher Christian, they never reached their destination. The exact cause of the mutiny is not completely clear to this day. Captain Bligh and eighteen faithful sailors were put on board a small boat and a month and a half later arrived safely in Timor. In 1792 Bligh remade the journey and this time did manage to transport trees to the West Indies. Such a story obviously should end with the breadfruit becoming an important food plant in the West Indies, but it never did. The West Indians did not eagerly adopt it, much preferring bananas and plantains and other foods already familiar to them. But as a testimony to Captain Bligh's persistence, the breadfruit is now well established in tropical America, the trees being appreciated for their ornamental value and occasionally used for food. Captain Bligh has been honored by having another tree named for him, *Blighia sapida*, the akee; its fruit is edible, but if eaten when unripe or overly ripe it can cause death.

Both seedless and seeded forms of breadfruit are known. Seedless breadfruit is generally prepared by boiling or baking. The seeded type is grown primarily for its seeds, called breadnuts, which are cooked and eaten. In parts of the Pacific area a cloth is made from the fibrous inner bark of the breadfruit tree. The details of the domestication of the breadfruit are unknown, but it has been suggested that the cultivated plant is of hybrid origin. The seedless forms obviously represent variants selected by people since they cannot propagate naturally.

The genus *Artocarpus* contains another species grown for its edible fruits and seeds. The jackfruit (*Artocarpus heterophyllus*) is native to the Malay region and widely distributed in the tropics today, although of less importance than the breadfruit. It, too, is an attractive tree, differing from the breadfruit in having entire rather than lobed leaves and much

Figure 8–7 Breadfruit tree. (Courtesy of Hawaii Visitors Bureau.)

Figure 8–8 Jackfruit. (Courtesy of USDA.)

larger, sweeter fruit. The fruit, which is reported at times to reach lengths of nearly three feet and weights of more than seventy-five pounds, has been stated to be the largest fruit of any cultivated plant. It may well be the largest fruit of any cultivated tree but there are pumpkins and squashes on record that far exceed it in size and weight. Moreover, these latter fruits each develop from the ovary of a single flower, whereas the fruit of jackfruit, like that of breadfruit, is made up of the ovaries from many individual flowers.

Bananas

Many people who think of the banana only as a dessert fruit may be surprised to find it included with the staples. In many parts of the tropics, particularly in East Africa, it is the principal food of various

Figure 8–9 Bananas. The pointed structure at the tip of the bunch (*bottom*) is the male bud. (Courtesy of USDA.)

peoples. Of the thirty-seven million tons of bananas produced annually, only about 15 percent enters the world trade, the remainder being consumed locally. Over one-half of the bananas are eaten raw in the way familiar to us, and about one-half are eaten cooked, as a vegetable. The dessert, or sweet, banana is sometimes cooked, but most cooking bananas are starchy rather than sweet and are referred to as plantains or cooking bananas.

The bananas had their origin in southeastern Asia, many in the Malay region. Our earliest record is an account from India in 500 B.C., but it is generally assumed that the banana is a much more ancient crop, although its exact age is unknown. Wild bananas have relatively small fruits with many hard seeds and probably were not a particularly attractive food. Other parts of the wild plants may have been eaten more frequently—the large underground part, or corm, the shoots, and the large male bud—as they still are in some places in the tropics. The leaf stalk also may have been used for its fibers. Thus, other parts of the plant were probably much more important to early people until there was a genetic change that led to seedless fruits. Seedlessness in bananas derives from both parthenocarpy, or the development of fruit without pollination, and sterility. The species that is the ancestor to our domesticated bananas is *Musa acuminata*, which still exists in the form of numerous southeastern Asian races. At some time this species hybridized with another, *Musa balbisiana*; today some of our cultivated bananas are "pure" *Musa acuminata* and others contain one or two chromosome sets from *Musa balbisiana*. An important event in the development of the edible bananas was the addition of a chromosome set to the normal diploid set. Such triploids, derived from crosses of diploids and tretraploids, are more productive and vigorous than diploid bananas and are also usually sterile and quite variable, giving us some superior plants to choose from. Today most of our bananas are triploids, although some diploids and a few tetraploids are cultivated.

The banana reached Africa at about the beginning of the Christian era, along with several other food plants from southeastern Asia. Some have thought that the introduction of these plants led to a population explosion in Africa. The plant was first heard of in Europe from a report of Alexander the Great, and Pliny wrote that it was the plant of the wise— hence one of Linnaeus' names for the banana, *Musa sapientum*, "of the wise men." Another Linnaean name formerly used was *Musa paradisiaca*, for the plant was thought to have been the Tree of Paradise or the Tree of Knowledge among some people. From Africa the banana was carried to the Americas in 1516 and became so well established in a short space of time that some early travelers thought that it was an indigenous American plant. From Africa, too, came the name *banana*.

Although the role of the banana in history may not be as great as that of the potato, it nevertheless has played a significant one in Central America, where many of the countries have become known as "banana republics." The story begins in 1870 when bananas were first exported from tropical America to the United States, where they proved to be immensely popular and profitable. In 1871 the American magnate, Minor Cooper Keith, built a railroad in Costa Rica. After looking for something for his railroad to carry, he began planting bananas there three years later. In 1899 the Keith interests merged with the Boston Fruit Company to form the United Fruit Company, which was to exercise control of the economics and governments of several Central American nations for more than fifty years. In fact, the overthrow of President Jacobo Arbenz of Guatemala in 1954 was initiated by the U.S. government to protect the interests of the United Fruit Company, although the reason given was to save the country from communism.

The United States imports more bananas than any other country. Some years ago Honduras and Costa Rica were among the leading exporters, but when a fungal disease destroyed many of the commercial plantings in Central America, the small South American republic of Ecuador became the world's chief exporter, a position that it holds today. The Central American countries are growing varieties more resistant to disease, and Honduras is now second in world production.

The banana not only comes in one of the neatest and most convenient packages of all of our food plants, but also is one of our best energy sources. Its nutritive value is very similar to that of the white potato, although in the dessert varieties more of the carbohydrate is in the form of sugar. The average person would need to eat about twenty-four bananas a day if they were the sole source of calories. Bananas contain considerable amounts of a substance called serotonin, which may be slightly poisonous, and Pirie has written that "dependence on them as the main source of energy would be inadvisable." But in parts of the world people do use them as their principal food source.

The banana plant, contrary to popular notion, is not a tree. The trunk of the banana is not woody and is not even a stem, but consists of the leaf stalks. It also may surprise some people to learn that the fruit is classed as a berry. The plant is a perennial herb, the aerial portions arising from a corm. The pseudostem produces a bunch of bananas and then is removed or eventually dies naturally, but sideshoots, or suckers, from the corm continue to grow, and stems from these will produce bunches of bananas in the following seasons. Suckers or corms are used for propagation. In a sense the plant is immortal, although in practice most banana plantations are started anew after five to twenty years. Some, however, have been known to remain in production for up to a hundred years.

The banana is for the most part a tropical crop. It needs considerable warmth and water, with adequate drainage. Some bananas, however, are produced in relatively dry subtropical areas, usually under irrigation. The plant needs little attention other than pruning to remove unwanted suckers. It is fairly taxing of the soil, and clean cultivation, or removal of all weeds in the plantation, which tends to promote erosion, is no longer recommended.

Only a very few of the some three hundred varieties of bananas ever reach the United States, and by far the most common in the past was the variety Gros Michel. The varieties known as Valery and Cavendish, more resistant to some diseases, have now largely replaced it. Other

Figure 8–10 Harvesting abaca. The trunk, which is made up of the leaf stalks, is used for its fiber. (Courtesy of USDA.)

varieties are known that are better tasting, but unfortunately they don't ship as well as those previously named. Bananas are cut while quite green, even for local use in the tropics, where the sweet types are allowed to ripen in a shady place near the house to be used as needed. Fruit intended for export is ripened under carefully controlled conditions and ethylene is sometimes used to hasten the process.

Although more than 90 percent of the bananas grown are used directly for food, several products are made from the remainder. A banana flour or powder is sometimes produced. According to the primitive method of drying the fruits for flour, the bananas were placed in heaps on mats over a mixture of cow dung and water and covered with leaves. Modern methods involve slicing the bananas and drying them by artificial means. Candies and various confections are made by splitting and drying the bananas. Sliced dried bananas are used as banana chips.

A beer for local consumption is made from bananas in parts of Africa. Other parts of the plant are sometimes used for food. The leaves are often employed for wrapping—the thick, waxy covering of the leaf makes it an ideal "wax paper"—or for plates, and at times for emergency umbrellas.

The leaf fiber of the banana is of no commercial importance, but members of other species of the same genus have valuable fibers. *Musa textilis,* commonly known as abaca, or Manila hemp, looks very much like a banana but has an inedible fruit; it produces an extremely strong fiber in the leaf stalk that is used to make high-grade cordage. The fiber is one of the principal exports of the Philippines. The plant became well established in tropical America during World War II and was grown widely in Central America on banana plantations devastated by disease.

Scientific breeding of the banana has been carried out for only half a century, nearly all of it in Trinidad. Much of the early work was necessarily concerned with learning as much as possible about the plant. Seeds are required for breeding studies, and fortunately most edible bananas will produce seeds, although in very small numbers, if pollinated. Thus far, there has been only limited success in producing improved varieties.

CHAPTER NINE

Coconut: The Tree of Life

No part of the coconut tree is wasted.

MALAYALAM PROVERB

Although most people in the temperature zone are aware that palms are graceful, attractive trees of the tropics, few realize their great usefulness. In fact, many botanists consider the palm family (Palmae) second only to the grass family in its importance to humankind, and in many parts of the tropics palms are far more important than the grasses. There are more than two thousand species of palms, and a long list of useful ones would certainly include the coconut, the date palm, the African oil palm, the rattan palm, the wax palm—carnauba, the world's preferred wax, comes from a palm—and the sago palm,* whose stem yields a starchy food used in Malaysia and Indonesia.

The date palm (*Phoenix dactylifera*) has been considered the "tree of life" in the subtropical deserts of the Old World and was a symbol of fecundity and fertility. Its fruit has long served nomadic Arabs as a staple food. In addition to its high sugar content (about 70 percent), the date contains about 2 percent protein and 2 percent fats, and is a fair source of some vitamins and minerals; it thus is a considerably better food than some of the starch crops. Like most palms, the date palm has many other uses—360 of them according to an ancient Persian source. Although the date palm is widely used, the coconut (*Cocos nucifera*) is neverthelelss the world's most important palm. Today the demand for its oil is not as great as it once was, but the coconut continues to serve people in many ways.

In addition to having been called "man's most useful tree," the coconut has also been referred to as "one of Nature's greatest gifts to man" and "mankind's greatest provider in the tropics." People use practically all parts of the plant in one way or another, but it is the fruit (botanically classed as a drupe, not a nut) that gives the plant its great economic importance. The fruit is made up of a smooth outer layer, a

---

*A number of plants share the common name sago palm, including several different species of the palm family; however, some plants known by this name are not members of the family.

Figure 9–1   Date palm in fruit, Israel. (Courtesy of FAO.)

Figure 9–2   Coconut tree in fruit. (Courtesy of Hawaii Visitors Bureau.)

Figure 9–3 Coconut. *Left to right:* Fruit as it comes from the tree; with outer rind removed to show coir; with coir removed, exposing shell (this is the way coconuts usually appear in markets in the United States); shell broken to expose the meat; coconut water.

fibrous middle layer, or coir, and a strong inner portion, or shell, that encloses the single seed. Frequently coconuts appearing in markets in the United States have had the outer layers of the fruit removed so that the shell is exposed. The seed, in addition to the embryo, has a thin, papery outer layer; inside this layer are the "meat," or kernel, and the "coconut water," which together form the endosperm, or reserve food for the embryo. The coconut seed is one of the largest known.*

Coconuts, like bananas, need plenty of warmth and moisture and good drainage. They are mostly found near the coast but do grow considerably inland in some regions. Florida is the farthest place from the equator where they are known to grow. Although many are now grown on plantations, they are still a crop on small holdings in many places. Ninety percent of all coconuts are grown in southeastern Asia. The Philippines and Indonesia now lead the world in their production. Mexico is the chief producer outside of southeastern Asia.

Coconut trees have been reported to bear five hundred nuts a year, but fifty to one hundred seems to be more normal. Harvesting is usually done by climbing the trees or by cutting the nuts with knives attached to long bamboo poles; more rarely the nuts are allowed to fall to the ground and are then collected. Monkeys have been trained to harvest coconuts in several countries of southeastern Asia. The name *Cocos* itself relates to monkeys but has nothing to do with their harvesting the fruit.

*The distinction of having the largest seed apparently belongs to another palm, *Lodoicea maldivica*, known as the double coconut, Seychelles nut, or *coco de mer*, which has a fruit two or three times the size of the coconut and weighing up to forty pounds. Marvelous tales were once told of the fruits that washed up on the shores of India from their homeland in the Seychelles Islands.

A

B

Figure 9–4
A. Monkey harvesting coconuts in Thailand. (Courtesy of Sakorn Trinandwan.) B. Coconut "monkey face."

The word, which comes from the Portuguese, means monkey's face, a reference to the three eyes in the shell that make it resemble the face of a monkey.

After the nuts are harvested they are cut in half, and the meat is gouged out immediately or after partial drying in the sun. The meat is then cured, by sun drying where weather permits or in kilns, to produce copra. The moisture content must be drastically reduced or the copra deteriorates rapidly. After the drying, the oil, which forms 60 to 70 percent of the copra, is extracted. Primitive methods using stone or wooden mortars and pestles, powered by humans or bullocks, are still employed in parts of India, but these have been replaced by hydraulic presses in most other coconut-producing areas. The oil is the most important commercial product derived from the coconut. Its greatest use in the last century was in the manufacture of soaps. All floating soaps were made from coconut oil until fairly recently, when it was found that soaps made from other oils would float if air was pumped through them during manufacture. Coconut oil is still regarded as one of the best for making soaps, and considerable amounts are still used for that purpose. In this century, however, its chief use has been for margarine, and it was the main oil used for that purpose until recently; soybean and cottonseed oils have come largely to replace it. Since coconut oil is composed of 90 percent saturated fatty acids, it is not recommended for human food as many of the other vegetable oils, which are largely unsaturated, are.

The residue, or coconut cake, left after oil extraction is used primarily to feed livestock. The coconut cake is a rich source of both protein—one of the most nearly complete proteins of all vegetable sources—and carbohydrates. Although it is more fibrous than most other oil-seed residues and hence difficult to digest, it is nevertheless unfortunate that more of it is not processed for human food, for coconuts are mostly grown in areas where protein deficiency is common.

The meat of either immature or mature coconuts is used directly for

Figure 9–5 Copra drying in Surinam. (Courtesy of FAO.)

human food in many areas where these palms are grown. Although it is the chief vegetable protein source for some people, it is usually served mixed with other foods as a vegetable, and nowhere does it appear to be a basic food staple as do most of the plants we have discussed. The per capita consumption is estimated to be 140 nuts a year in Sri Lanka and is probably considerably greater than this among Polynesians in some of the Pacific Islands. In many countries of the temperate zones dried coconut is used mostly for candies and cakes. Desiccated coconut meat was first made in England and the United States at the end of the last century and is a fairly important use of coconut meat today.

The fibrous part of the fruit, the husk, or coir, also has a number of uses. It makes a fine rope, resistant to sea water, and its use for this purpose is quite ancient. Coir is also used to make mats, rugs, and filters, and as stuffing for furniture. India is the world's greatest producer of coir. To prepare the fibers, the husks are immersed in saline backwaters for several months for retting. The fibers are then separated by beating the husks with wooden mallets or clubs.

The young inflorescence, or flower cluster, of the coconut, like that of several other palms, yields a sweet juice, or toddy, when tapped. The toddy, which is mostly sucrose, is sometimes drunk fresh; more frequently it is used to prepare alcoholic beverages, such as arrack, or vinegar. In Sri Lanka more than eight million gallons of arrack are produced annually, very little of it being exported. Small amounts of toddy are also used for making sugar.

The shells have a number of uses, the most ancient of which are as eating or drinking utensils and for fuel. They are still used for these purposes as well as for bowls for hookah pipes and for the manufacture of novelties or "artistic objects"; when ground they are used as a filler in plastics. The coconut water makes a refreshing drink, and in recent years has been used by plant physiologists as a growth-promoting substance. The large leaves, which reach lengths of twelve feet, are used for thatching and for making baskets and hats.* The wood is used to some extent in construction and furniture making, and forms some of the "porcupine wood" of commerce. The palm heart, or "cabbage," the tender bud at the apex of the stem, is sometimes eaten, but not as frequently as that of some other palms, for once the bud is removed the tree dies. The coconut is also still used as a religious offering in some parts of southeastern Asia, probably stemming from the ancient belief

---

*They are apparently second only to the leaves of the Panama hat palm for this purpose. The Panama hat palm isn't a true palm but, rather, belongs to the family Cyclanthaceae, and most of the hats come from Ecuador.

Figure 9–6 Girls gathering coconuts in Sri Lanka. (Courtesy of FAO.)

that the coconut is the "Tree of Heaven"—*Kalpa Vrikska* in India—or the "Tree of Life." There is also a belief among certain people of New Guinea that the first coconut tree sprang from the head of the first person to die.

Two main groups of coconut palms are recognized—dwarf forms and tall forms. Numerous varieties exist within each group, differing primarily in the shape, size, coloring, and yield of the fruit. Although seed selection for high yield probably is fairly ancient, there has been very little modern scientific breeding work on the coconut. There are several reasons for this. First of all, it is a tropical crop, and as we have already seen for other plants, these have received much less attention than the crops of the temperate zone, where more plant breeders and money for research are available. Moreover, the coconut is often a small landholder's crop rather than a plantation crop, and governments seldom subsidize small farms as much as big ones. Nor can small landholders employ breeders, as do some of the companies that have extensive holdings. Another factor responsible for the limited improvement of the coconut is that breeding work with a tree always takes longer than that with annual or herbaceous perennial plants. It takes three years for a dwarf coconut palm to bear fruit and five to seven years for the tall varieties. From time of flowering to fruit maturity is nearly a year. Germination of the seed requires about four months, time enough for a full growth cycle in some annual crops. Thus, the breeding of superior varieties of the coconut through hybridization requires many years of work.

The place of origin of the coconut has been the subject of some controversy in the past. There were some who held that it was a native of the New World, primarily because all of its close relatives are American. Others maintained that the coconut was native to the Indo-Pacific region, pointing out that it was a much more extensively used plant in that area than in the Americas. While it is true that the coconut does have most of its close relatives among the American palms, there is now general agreement among botanists that it originated in the Indo-Pacific region. Among the most convincing evidence that has been brought forward since the original controversy is the discovery of fossil *Cocos* of late Tertiary age in New Zealand and India, proving that a species of coconut did inhabit the Pacific region before modern humans appeared on the scene.

The coconut was present in both Asia and the Americas previous to 1492. At that time, however, it was known only from western Panama in the Americas, and its wide distribution in the New World came about in historical times. Its presence in Panama previous to the arrival of the Spanish can probably be explained without invoking human aid, for coconuts are able to float in sea water for more than a hundred days, which would allow ample time for a fruit to float across the Pacific. Probably the establishment of coconuts in new areas through the agency of ocean currents is a rare event, but it would need to have happened only once to explain the plant's presence in the New World. People, as well as ocean currents, are probably responsible for its wide distribution on the Pacific Islands. How early the coconut first became a domesticated plant is not known. It was grown in India by 1000 B.C. but it may not have been first domesticated there.

Two tales concerning coconuts have been widely circulated. One is the story of the coconut, or robber, crab. According to the account given by Darwin, this crab tears the husk from the coconut "fibre by fibre," exposing the three eyes on the shell, and then hammers on one of the eyes with its heavy claw until a hole is made, after which it extracts the coconut meat with its pincers. Apparently Darwin's leg was being pulled by a Mr. Leish, whom he credits as his source, for Reginald Child (1974) points out that no one has ever seen a crab in nature perform this remarkable feat, and in feeding experiments the crabs have died when coconuts were the only food offered them. The second story, concerning the finding of pearls inside coconuts, also has never been substantiated, according to Child. The so-called coconut pearls in museums have been shown to come from mollusks. There is, however, a recent account* in which it is claimed that there is scientific evidence that on very rare occasions "pearls" do develop in coconuts.

*J. Baltes, "Ueber die Kokosperle," *Fette Seifen Anstrichmittel* 73 (1971): 1–4.

CHAPTER TEN

Oils: Sunflower and Cotton

Everything is soothed by oil.

PLINY THE ELDER, *Natural History, Book II*

Cotton is King.

DAVID CHRISTY, 1855

Fats and oils not only provide a more concentrated source of energy than carbohydrates and proteins but also are a necessary part of the human diet. In recent years there has been a great increase in the demand for oils of vegetable origin, particularly for making shortening and margarine; the latter has largely replaced butter in many parts of the world. The reasons are twofold: vegetable oils are usually cheaper than animal oils; and animal fats and oils contain saturated fatty acids that may be harmful to our health, whereas many plants contain polyunsaturated fatty acids that may be beneficial. A number of plants furnish edible oils, the soybean being the world's chief source. Other plants contributing large amounts are sunflowers, peanuts, palms—chiefly the coconut and the African oil palm—rape (see Chapter 11), cotton, and olive. Corn oil is of considerable importance in the United States for cooking and for making margarine. In addition to their use for food, many plant oils have considerable use in industry.

One of the newest plant oils comes from the jojoba (*Simmondsia chinensis*), a wild shrub of the southwestern United States and northwestern Mexico. This oil has been found to differ from all other oils—chemists refer to it as a liquid wax—and its unique chemical structure gives it a number of potential uses. Presently its greatest use is in cosmetics, and it holds great promise for giving high-quality lubricants that can substitute for sperm-whale oil. It also may serve as an edible oil but a great deal more research is necessary to prove its safety and effectiveness. The meal left after the oil is removed is high in protein, but the presence of toxic factors has thus far not permitted its use for animal feed. The jojoba has been under cultivation in the arid parts of the United States as well as in many other countries for fifteen years, and it is now on its way to becoming a domesticated plant.

Figure 10–1 Harvesting olives in Tuscany, Italy. (Courtesy of FAO.)

The olive (*Olea europaea*), a tree native to the Mediterranean region, is one of the most ancient sources of oil for human use. It is still one of the world's major oil sources, but its importance has greatly declined in recent years with the development of other plant oils. Most of the world's production is centered in its original homeland; Italy is the leading producer. The tree is also cultivated in California where it was introduced in 1769. Well over 90 percent of the crop goes to the production of oil. It is unusual among the oil crops in that the oil comes from the fruit rather than the seed.

Sunflower

The sunflower (*Helianthus annuus*) is the only important crop plant to have been domesticated in what is now the United States. The wild sunflower was an important food source to seed gatherers in western North America, and people apparently carried it as a weed to the central United States, where it became domesticated sometime before 1000 B.C.

Native Americans esteemed the sunflower seeds as food and a source of oil; they also found many other uses for the plant. When the Europeans arrived they found the sunflower fairly widely cultivated in eastern North America as well as in the Southwest and in northern Mexico, although it was nowhere a major food plant. After its introduction into Europe in the sixteenth century, the sunflower was at first regarded as a curiosity, chiefly for the great size the plant could attain in a single season; it did not become a major food plant until it reached Russia. Large areas in Russia were found suitable for growing sunflowers, which soon became the Russians' major source of food oil. For many years the Soviet Union has led the world in sunflower production. Much of the modern improvement of the sunflower was done in Russia, including an increase in oil content from 28 percent to around 50 percent and the development of semidwarf varieties that could be harvested mechanically. The sunflower became an important crop in the Balkans, a little later in Argentina, and more recently in several other countries. Although rather widely grown on a minor scale in the United States for many years, chiefly for bird seed, it has become important as an oil plant in this country only within the last twenty years. A significant factor in the recent success of the sunflower as a major oil source was the development of high-yielding hybrids using a cytoplasmic male-sterile line.

Figure 10–2 Achenes of sunflower from archaeological site in Ohio (*left*) and from modern variety. The archaeological achenes are charred and are probably 10 percent smaller than they were originally. (From *The Sunflower* by Charles B. Heiser, Jr. Copyright © 1976 by the University of Oklahoma Press.)

Figure 10–3 The sunflower increased greatly in size with domestication. Wild sunflower (*lower left*); weed sunflower (*upper left*); and domesticated sunflower. (From *The Sunflower* by Charles B. Heiser, Jr. Copyright © 1976 by the University of Oklahoma Press.)

Although the sunflower is cultivated today chiefly for its fine oil, fair amounts are still grown to supply seeds for confections such as "sunflower nuts" and for feeding birds. The seed cake left after the removal of the oil is a protein-rich food used for feeding livestock.

The cultivated sunflower is an annual plant that produces a single stalk bearing a massive head with numerous small flowers. Each flower has an ovary that produces a dry fruit known as an achene, which contains a single seed. Wild sunflowers, in contrast, are branched plants that bear numerous heads; they also produce much smaller achenes. The Native Americans had produced a domesticated plant close to the quality of modern varieties, except for the yield of oil, long before the arrival of the Europeans.

The sunflower is a member of the plant family Compositae, which is the second largest family of the flowering plants. Only the orchid family is larger. In spite of its size the family has furnished few other food plants, lettuce (*Lactuca sativa*) perhaps being the most important. The Jerusalem artichoke, or "sunchoke" (*Helianthus tuberosus*), grown for its edible tuber, is a perennial sunflower and, like the sunflower itself, was domesticated in temperate North America. The safflower (*Carthamus tinctorius*), of Old World origin, is another member of this family. Apparently originally domesticated for its flowers, which were used for a dye, it was subsequently grown for the oil content of its seeds. Because of its high content of polyunsaturated fatty acids, safflower oil is considered one of the finest for human consumption. Although the family Compositae has supplied few food plants, it more than makes up for this with a large number of very showy ornamentals, including marigolds, zinnias, chrysanthemums, and various forms of the sunflower.

Cotton

Although cotton is a very old domesticated plant, its use as an oil crop is only a little over a century old. Still the world's most important vegetable fiber, producing more fiber for clothing than all other textiles combined, it now also ranks as one of the major sources of vegetable oil and stock feed, both derived from its seed. The fiber, called lint, comes from hairs on the seed. Before the seeds were used for oil those not needed

Figure 10–4 Cotton plant in fruit. (Courtesy of USDA.)

Figure 10–5   Harvesting cotton by hand in Yugoslavia. (Courtesy of FAO.)

for planting were discarded. For every 500-pound bale of cotton, nine hundred pounds of seeds are available for oil, seed meal, and other uses.

The several species of cotton were originally shrubby or sometimes treelike perennials, but under domestication annuals developed. The rather large flowers produce a capsular fruit known as a boll. Cotton is a member of the Malvaceae, or mallow, family, which is better known for its ornamentals, such as hollyhock and *Hibiscus,* than for food plants. Okra, or gumbo (*Abelmoschus esculentus*), of Old World origin, is one of the few species now fairly widely grown as a vegetable.

Four different species of cotton have been brought into cultivation. *Gossypium herbaceum* was domesticated in Africa, and *Gossypium arboreum,* probably derived from it, was grown in India. Cloth fragments from India dated at around 3000 B.C. indicate considerable antiquity for the Old World cottons. In the New World two species were also brought under domestication, *Gossypium hirsutum* in Mexico and *Gossypium barbadense* in western South America. There is archaeological evidence for the existence of *Gossypium hirsutum* around 2500 B.C. and for *Gossypium barbadense* about a thousand years later. A study of the chromosomes of these species revealed that the Old World cottons were diploid and the

New World cottons tetraploid. It was further found that the New World cottons had one chromosome set, designated AA, identical to that of the Old World cottons, and a second set, designated DD, that was different. Thus it was apparent that the New World cottons owed their origin to hybridization between species with the AA and DD sets followed by chromosome doubling. Wild diploid species of *Gossypium* exist in the Americas with the DD set, but no species with the AA set occur there. When this was learned some people postulated that humans must have been responsible for carrying an AA species from the Old World to the New to allow for the hybridization to take place. More recently, however, this hypothesis has been largely replaced by one that has the Old World species arriving in the Americas by natural means, perhaps by seed drifting across the Atlantic. It is now thought that the New World tetraploid cottons developed before humans appeared on the scene, or when humankind was still in its infancy.

The American cottons are superior to those of the Old World and gradually largely replaced them. *Gossypium hirsutum*, earlier maturing than *Gossypium barbadense* and hence more resistant to boll weevil, is said to furnish 95 percent of the cotton grown in the world today. China, the Soviet Union, and the United States are the world's leading producers. The southwestern United States, where the crop is grown under irriga-

Figure 10–6 Harvesting cotton in Arizona. A mechanical harvester can pick as much in an hour as a person could pick in seventy-two hours. (Courtesy of USDA.)

Figure 10–7   A harvest of cotton in Upper Volta. (Courtesy of FAO.)

tion, has replaced the southeastern states where cotton was once king as the chief area of production in the United States.

The boll weevil, a beetle commemorated in song, entered the United States from Mexico late in the last century and has proved to be the most serious pest of cotton. Losses to the crop have been estimated at over $200 million a year. Breeding work devoted to securing resistant types and varieties that mature before the beetle has completed its life cycle has proven valuable, but the boll weevil continues to be a problem in some places. Large amounts of chemical herbicides, insecticides, and defoliants (used to remove the leaves to make harvesting easier) are used on the cotton crop and have raised concern over the resulting environmental pollution.

Cottonseed oil is used mostly for the production of shortening and margarine. The seed meal that remains after the extraction of the oil is a protein-rich food, at present used mostly for animal feed but potentially of greater use for humans. In 1949 the Institute of Nutrition of Central America and Panama (INCAP) was formed, and one of its significant achievements has been the development of a vegetable flour, incaparina (from INCAP and *harina*, the Spanish word for flour), made largely from cottonseed meal. Incaparina is 25 percent or more protein. Its creation was an attempt to improve the diet of Latin Americans, who subsist mostly on maize and hence often suffer from protein deficiency. Incaparina is much cheaper than milk and has about the same nutritional value.

A problem in the use of cotton products for food is that the plants have glands that produce a toxic substance, gossopol, which must be removed before its products can be eaten by nonruminant animals. Gossopol is now removed chemically, but the process is expensive and reduces the value of the protein. Plant breeders have succeeded in producing glandless plants, but these are more susceptible to insect damage than the plants with glands.

In the 1950s it was discovered in parts of China that people whose food was cooked in cottonseed oil showed considerable infertility, particularly the men. Research was begun and in 1979 the Chinese announced the development of a male contraceptive, called gossopol, from cotton seed. The drug is very effective as a contraceptive but has some undesirable side effects. Research with it continues.

CHAPTER ELEVEN

To Complete the Meal

Cauliflower is nothing but cabbage with a college education.
MARK TWAIN, *Pudd'nhead Wilson's Calendar*

Although we can, and sometimes do, live solely on the basic foods, we usually use other plants as food or beverage with every meal. Some of these other foods are eaten primarily to add variety to the diet and to increase the enjoyment of eating, but at the same time many of them are excellent sources of vitamins and minerals and also supply small amounts of carbohydrates, proteins, and fats. Primitive people, as pointed out in the first chapter, almost certainly exploited a great array of wild food plants, and they continued to use some wild plants as fruits and vegetables after the domestication of the cereals. Eventually some of the fruits and vegetables were cultivated, and in time they became completely domesticated species. People today still utilize wild food sources in many parts of the world, often out of necessity but sometimes by preference. The collecting of wild foods has become an interesting hobby for many people in the United States.*

Vegetables

The word "vegetable" has no precise botanical meaning in reference to food plants, and we find that almost all parts of plants have been employed as vegetables—roots (carrot and beet), stems (Irish potato and asparagus), leaves (spinach and lettuce), leaf stalk (celery and Swiss chard), bracts (globe artichoke), flower stalks and buds (broccoli and cauliflower), fruits (tomato and squash), seeds (beans), and even the petals (Yucca and pumpkin). A great many different plant families have provided our vegetables, but just as certain families have been particu-

---

*Many of the "wild" foods being gathered in the United States are introduced species that have escaped from cultivation rather than native wild plants. Wild asparagus, wild carrots (also known as Queen Anne's lace), prickly lettuce, chicory, and burdock, for example, are plants that were introduced, either intentionally or accidentally, from the Old World and have become widespread weeds in North America. The Jerusalem artichoke and cattail, also used as wild food sources, are native North American species.

larly important in giving us our staples, so too have certain others in giving us vegetables.

The mustard family, or Cruciferae, has been particularly significant for its vegetables, and a single species, *Brassica oleracea*, has provided us with the cole crops, which include cabbage, kale, brussels sprouts, cauliflower, broccoli, and kohlrabi. The kohlrabi, although often cultivated in the United States, is seldom seen in markets. In this variety, the stem enlarges above ground level to produce an edible tuber. The ancestor of all the cole crops was native to the Mediterranean region, and some think that it was first used for its oily seeds. Selection with emphasis on different parts of the plant eventually produced the diversity of cultivated varieties we now enjoy.

Other species of *Brassica*, originally native to either Europe or Asia, account for a number of other vegetables. These include plants grown for their roots—the turnip and rutabaga—and a great number whose leaves are eaten, collectively designated as the mustards. Two of these, brown mustard and black mustard, have seeds that are widely used in the preparation of the condiment mustard, which after black pepper is the most widely used spice in the United States. Mustard is also used in medicine as a rubefacient or counterirritant. The seeds of two species (*Brassica campestris* and *Brassica napus*) called rape—a word derived from *rapum*, Latin for turnip—have long furnished an industrial and edible oil. The presence of erucic acid in the seeds, however, was thought to represent a health hazard in human consumption. Plants have been developed that lack or have very low concentrations of the acid, and since 1975 they have provided one of the world's most widely used edible oils. Its low content of saturated fats makes the oil one of the most desirable ones. Rapeseed is produced mainly in Canada, China, India, and parts of Europe. The oil is known as canola in North America. The radish (*Raphanus sativus*) is another Old World member of the mustard family employed for food. Although radishes are ordinarily used only as a salad ingredient or relish in most places, they have an important role as a food plant in the Orient. Japanese varieties may reach weights of sixty-five pounds. They are used as a cooked vegetable, often stored for use in the winter, and are also fed to livestock. In parts of Asia, one variety of radish is grown especially for its seed pods, which may reach lengths of two feet and are used as a vegetable.

Of equal or greater importance for its contributions to our vegetables, is the cucurbit family (Cucurbitaceae). Five different species of squash or pumpkin belonging to the genus *Cucurbita* were domesticated in the Americas. Some of these rank among the oldest known foods of the Americas, being recorded in archaeological deposits from 7000 B.C. in

Wild type

Kale

Cabbage

Cauliflower

Brussels sprouts

Broccoli

Kohlrabi

Figure 11–1 Variation in *Brassica oleracea*. Human selection has produced varieties valued for their leaves (kale and cabbage), specialized buds (brussels sprouts), flowering shoots (cauliflower and broccoli), and enlarged stems (kohlrabi).

Mexico. Since the wild cucurbits have little or no flesh in the fruit, it has been postulated that they may have been domesticated for their edible seeds—"pepitas," or pumpkin seeds, are still eaten. Mutant types with fleshy fruits then appeared, according to the theory, and their deliberate selection has produced the thick-fleshed varieties now widely cultivated. Following the European discovery of America, pumpkins and squashes were soon introduced into Europe and Asia, and today they are important in many parts of the world not only for human food but for livestock as well. The largest fruit in the plant kingdom comes from a squash (*Cucurbita maxima*); some weighing nearly seven hundred pounds have been reported. The Old World has also furnished food plants from the Cucurbitaceae, including the cucumber, the melons such as cantaloupe and cassaba, and the watermelon. The cucumber and the melons come from different species of the genus *Cucumis;* the watermelon belongs to the genus *Citrullus.*

The bottle gourd (*Lagenaria siceraria*), another member of the Cucurbitaceae, has never been more than a minor food plant, but has been valued for its hard-shelled fruits, which have been used as containers, for musical instruments, and for floats, as well as in other ways. This species is thought to be native to Africa, but archaeological remains of the fruit have been found in both Peru and Mexico dated at 7000 B.C. or earlier, and from the historical record we know that it is a very old

Figure 11–2   A harvest of bottle gourds, showing variation in the fruits. (From *The Gourd Book* by Charles B. Heiser, Jr. Copyright © 1979 by the University of Oklahoma Press.)

cultivated plant of both India and China. Some have thought that people may have been responsible for its wide dispersal in early times, but since it has been shown that the gourds can remain in sea water for long periods without damage to the seeds, it is perhaps more likely that its wide distribution is to be explained as the result of oceanic drifting of the fruits. It was probably the most widely distributed domesticated species in prehistoric times and continues to be fairly widely used throughout much of the tropics.

The nightshade family, or Solanaceae, in addition to providing the Irish potato has supplied us with several other food plants, including the tomato (*Lycopersicon esculentum*), which has become one of the world's most important vegetables. Although some books give Peru as its place of origin, the tomato became domesticated in Mexico. Its wild ancestor, similar to our cherry tomatoes, however, is native to Peru. The tomato was already a well established cultivated plant in Mexico when the Spanish arrived. It reached Europe in the first half of the sixteenth century and somehow acquired the reputation of being harmful to eat for reasons that are not entirely clear. It seems likely that it was recognized as a member of the nightshade family, known to Europeans of the time as comprising only poisonous plants such as deadly nightshade, henbane, and mandrake, and that people were therefore reluctant to eat it.

Although the tomato is regarded as a vegetable, the part that is eaten is, botanically speaking, the fruit. Tomatoes are known with yellow, orange, pink, and green fruits in addition to the familiar red types. One of the first tomatoes to reach Italy was a yellow-fruited variety called *pomi d'oro* (apple of gold), which somehow was transformed to *poma amoris* (apple of love). The name love apple soon became attached to the tomato, not because of any real or supposed aphrodisiac property but simply through translation of the transformed Italian name. Only in this century did the tomato finally become widely appreciated for the fine food that it is. Remarkable changes have been achieved by plant breeders in recent years, one of which is the development of tomatoes with special characteristics that allow them to be mechanically harvested. However, some of the varieties developed for long-distance winter shipping to northern markets are inferior in taste to those grown in the home garden. A biotechnology firm has recently developed a technique through genetic engineering that should allow better-tasting tomatoes to reach our markets.

Among other solanaceous plants grown for their fruits are the eggplant and the sweet and hot, or red, peppers. The eggplant, a species of *Solanum*, apparently had its origin in India and like the tomato was

Figure 11–3 Tomatoes that are tough skinned, even-ripening, readily detachable, and of a uniform size have been developed for mechanical harvesting. A harvester is in the background. (Courtesy of USDA.)

regarded with suspicion when it first reached Europe. One name for it at that time was mad apple, for it was thought that the eating of eggplant would produce insanity. Several different peppers, species of *Capsicum*, were domesticated in tropical America for their pungent fruits and became almost indispensable in the diet of many Native Americans. In post-Columbian times peppers became widely dispersed, and they have become as important in parts of southeastern Asia and Africa as they are in their homeland. In the temperate zones sweet peppers, a variety of *Capsicum annuum* (Figure 12–1), which also includes cayenne and chili peppers, have become more important than the pungent forms.

Because of its great contributions to our food supply as well as our ornamentals (for example, the petunia, *Petunia hybrida*), the Solanaceae must be considered a family of major importance. However, one member of the family has definitely not contributed to human welfare. Tobacco (*Nicotiana tabacum*), which was domesticated in South America, spread around the world with great rapidity after the discovery of the New World, and it continues to be one of the world's major economic plants in spite of the wide recognition of its harmful effects.

The parsley family (Umbelliferae), has provided several important vegetables as well as many of our spices. In addition to parsley (*Petroselinum crispum*), whose leaves are used in cooking and for garnishing and whose roots are sometimes eaten, the roots of the carrot (*Daucus carota*), the parsnip (*Pastinaca sativa*), and the leaf stalk of celery (*Apium graveolens*), all plants of Old World origin, are widely used vegetables. Celeriac is a variety of celery grown for its large edible root. The *arracacha* (*Arracacia xanthorrhiza*), or *zanahoria blanca* ("white carrot"), is another excellent vegetable whose use is little known outside of its homeland in the Andes. To this family also belongs the deadly plant called poison hemlock, native to the Old World but widely naturalizd in the United States.

Known better for its ornamentals than food plants, the lily family (Liliaceae) has also furnished us with some of the most ancient and still widely appreciated vegetables—onions, garlic, leeks, and chives. All of these, and some lesser known ones, belong to different species of the genus *Allium*. Several of them had early origins in the Near East and well-preserved remains have been found, particularly in tombs in Egypt. The edible parts are the bulbs, composed of leaf bases, or the leaf blades themselves. In garlic the bulb is made up of several smaller bulbs called cloves, which are used for propagation. Most of the other species are grown from seed. The distinctive odor of these plants, which makes them very appealing to some people but certainly not to all, is provided by sulphur-based compounds. Onions are an important item in international trade, and the Netherlands is the leading exporting country.

Although the avocado (*Persea americana*), once called alligator pear in the United States, is one of the most ancient of the domesticated trees of the Americas, only relatively recently has it become widely accepted and appreciated in the United States. The fruit consists of a fleshy pulp surrounding a single large seed. In contrast to most of our fruits and vegetables, the pulp, which is the part eaten, has a high content of oil—up to 30 percent in some varieties. Avocado seeds are found in archaeological deposits in Mexico that go back several thousand years; there are also early reports from Peru. In most of Latin America avocado is know as *aguacate*, the Mexican name, but in Peru it is called *palta*. The greatest production occurs in Mexico; the United States is now second, where it is cultivated in both California and Florida. The avocado is a member of the Lauraceae, a family mostly of aromatic trees of the tropics and subtropics. To the same family belong the laurel, cinnamon, camphor, and sassafras.

Fruits

Many fruits, wild or cultivated, must have always been a source of pleasure because of their sweetness. Like vegetables, they come from many different families of plants, but in the north temperate zone one family—the rose family, or Rosaceae—stands out. Among its more important contributions are the apple and the pear, or pome fruits, species native to West Europe and Asia; various species of stone fruits, genus *Prunus*, including the peach, cherry, plum, and apricot, most of which also come from the Old World (although species of cherry and plum were also domesticated in the Americas); and the "berry"* fruits, blackberry, raspberry, and strawberry, with domesticated species from both the New World and the Old. Although these are rightly called temperate-zone fruits, some of them are cultivated at high altitudes near the equator.

More and more of the tropical and subtropical fruits are now reaching the temperate zone markets. The citrus fruits were among the first and, as is well known, have become extremely important crops in the frost-free parts of the United States. All citrus fruits are members of the family Rutaceae, and a single genus, *Citrus*, supplies us with sweet, bitter or sour, and mandarin oranges (one form of the last is known as the tangerine in the United States), lemon, lime, grapefruit, citron, and shaddock, or pumelo. All of these originated in southeastern Asia, with the exception of the grapefruit, which is thought to have been derived from the shaddock after the latter was introduced in the West Indies. The citrus plant is a small tree, often somewhat spiny, and has attractive, fragrant flowers. The fruit is a special type of berry known as a

*Most of the so-called berry fruits do not have fruits that are classified as berries according to the botanical definition. A true berry is defined as a fleshy, many-seeded, indehiscent fruit that develops from the ovary of a single flower. Examples are grape, tomato, pumpkin, and orange. The raspberry and blackberry are in reality aggregate fruits in that they develop from many ovaries of a single flower. The individual fruitlets, or seed-bearing structures, of a blackberry, for example, are each the product of a single ovary and each is the equivalent of a plum in that it has a single seed enclosed in a fleshy covering. The strawberry is defined as an accessory fruit, for the fleshy part develops from a structure other than the ovary. The true fruits of a strawberry are the small, hard, straw-colored "seeds" on the surface.

hesperidium. Its thick, leathery rind bears numerous oil glands that yield an essential oil widely used in flavoring.

The nutritional value of citrus fruits is widely recognized today, and orange juice to start breakfast has become traditional in many parts of the world. Scientific proof of the importance of the citrus fruits came in 1756, when John Lind, a surgeon in the English navy, found that the scurvy common among sailors at the time could be prevented by eating oranges and lemons. Later in the century the Royal Navy began to provide rations of lime or lemon juice to its men, and the name "limey" came into use for British sailors as a result. Not until 1933 was vitamin C (ascorbic acid) identified as the factor responsible for the prevention of scurvy.

Considered by many to be the most delicious of all fruits, the mango (*Mangifera indica*) spread from India to southestern Asia in pre-Christian times and much later reached Africa and the New World tropics. The plant is a tree, sometimes attaining large size, and the succulent pulp of

Figure 11–4 Pineapples in Taiwan. (Courtsy of FAO.)

the fruit encloses a single large seed. The world's greatest production is in Asia. The tree is grown in California and Florida but not on a commercial scale. It is a member of the family Anacardiaceae, which also contains the cashew and pistachio as well as poison ivy. Unfortunately, some people have allergic reactions not only to poison ivy but also to other members of the family, including the fruit of the mango.

The pineapple (*Ananas comosus*) was encountered by Columbus on his voyages to the Americas, and it was not long before it spread throughout the tropics of the Old World. The pineapple is a member of the Bromeliaceae, a family much appreciated for its hothouse ornamentals. Most members of the family are epiphytes—plants that grow upon other plants but are not parasitic. The pineapple, however, is terrestrial. The plants are two to four feet tall with whorls of sword-shaped, sawtooth-edged leaves. The part that is eaten develops from the ovaries of many flowers, like the breadfruit, and thus is classified as a multiple fruit. Most commercial varieties are seedless, hence propagation is vegetative. Some pineapples, however, do produce seeds. Pineapples became a crop plant in Hawaii in the early part of the nineteenth century and Hawaii continues to be the leading exporter. Brazil, the original homeland of the pineapple, however, leads the world in pineapple pro-

Figure 11–5   Papaya. (Courtesy of FAO.)

duction. Pineapples contain a digestive enzyme, bromelain, and as some people have found to their dismay the raw fruit can not be added to gelatin if one wants it to gel. Native Americans used the pineapple not only as an aid to digestion but also in the curing of wounds. Stems are used to produce commercial preparations of the enzyme, which may prove to have significant uses in medicine today.

Much esteemed in the tropics, the papaya (*Carica papaya*) is still little known in most of the United States. The fruits that appear in the midwestern markets, usually less than half a foot long, come from Hawaii. Fruits three to four times that size, however, are common in tropical America, the homeland of the papaya. The latex from the fruit yields an enzyme, papain, that is widely used in meat tenderizers.

One of the newest fruits in our markets, and now rather common, is the kiwi berry (*Actinidia chinensis*). It was earlier called the Chinese gooseberry. Although not related to the fruits that are usually called gooseberry, it is native to China. The plant may have been domesticated in China quite early, but the kiwis we eat today stem from the domestication of the wild plant in New Zealand at the beginning of this century. Most of the kiwis in our markets are imported from there, although they are also grown commercially in California.

Nuts

Nuts of various kinds have long served as a highly concentrated source of nutrition. From the archaeological record we know that nuts of various wild plants were a frequent source of food in prehistoric times. The word nut, as popularly used, is applied to the fruit or seed of a great number of plants, mostly trees. Botanically, a nut is defined as a hard and indehiscent one-seeded fruit; of the "nuts" utilized by various peoples, only a few, such as the acorn, the chestnut, and the hazelnut, meet the botanical definition. Acorns from various species of oaks, in both the Old World and the New, were at one time an important source of food. Many of the Native American tribes of the west coast of North America relied on acorns as the principal food source, and they devised various ways of leaching the tannins and bitter principles from them in order to make them palatable. Acorns are still sometimes used as a food source by the poorer people in some of the Mediterranean countries of Europe. The Eurasian chestnut continues to be a food plant in southern Europe, but the American chestnut, whose nuts were once widely sought, has been practically eliminated by a blight disease that swept through the eastern United States at the beginning of this century. Acorns and chestnuts are valued as foods because of their high carbohydrate content.

Among the nuts with a particularly high protein content are the almond and the pistachio, both old cultivated plants of the Mediterranean region. The almond belongs to the same genus as the stone fruits, but the fleshy covering is poorly developed and the seed is the only part eaten. The almonds produced in the United States are grown in California.

Nuts with a high oil content include the Brazil nut and the cashew, both native to Brazil, the pecan of the central and southern United States, walnuts, and hazelnuts. The walnut foremost in use for food comes from the English, or Persian, walnut tree, which originally came from Iran. California is today one of the leading areas for walnut production. The native American black walnut is valued more for its wood than its nuts. The hazelnut, or filbert, of Europe is also an important yielder of nuts. There are also native American species of hazelnuts.

The macadamia (*Macadamia integrifolia*), the most recent nut to achieve popularity, is by many already considered the aristocrat of the nuts. The macadamia tree, native to Australia and the only plant of that continent to be developed as a commercial food plant, was introduced to Hawaii late in the last century. Its development as a crop was slow but it is now Hawaii's third largest argicultural commodity, exceeded only by pineapples and sugar cane. The tree is grown in many parts of the world today, including California and Florida, but more than 90 percent of the world's production still comes from Hawaii.

Beverages

It was discovered early that parts of certain plants had a pleasant stimulating effect,* and today many of these plants serve as the chief sources of nonalcoholic beverages. The four most significant of these are coffee, tea, cacao, or chocolate, and maté, or Paraguay tea. These plants, all members of different botanical families, share one important feature, the possession of caffeine or a very similar alkaloid that is responsible for their stimulating property. Except for cacao, these plants offer little or nothing in the way of nutrition and hence are hardly essential. Even though coffee and tea are not, strickly speaking, food plants, they are extremely important export crops in many parts of the tropical world and figure prominently in the economic welfare of many countries.

From a commercial standpoint coffee is the world's foremost beverage plant, although more people drink tea. Native to Ethiopia, coffee was

---

*Exactly how the effects of certain of these plants were discovered is somewhat of a puzzle, for some of them have little or no stimulating action unless they are specially cured or processed.

Figure 11–6   Coffee berries. (Courtesy of FAO.)

carried to Arabia over five hundred years ago, and for two centuries Arabia was the principal producer. Coffee was later found to be well adapted to many parts of the American tropics, from elevations near sea level to six thousand feet, and today Brazil, followed by Colombia, leads in the world's production. The United States is the world's chief importer of coffee, but per capita consumption is said to be greater in Sweden. The coffee plant (*Coffea arabica*) is a small tree or shrub with shiny, dark green leaves and numerous clusters of white flowers. Each of the berries contains two seeds called coffee beans. Following harvest, a process of fermentation and roasting is required before the beans assume their distinctive odor and flavor. Two other species of *Coffea* are cultivated, but they furnish only about 10 percent of the world's supply. One of them, *Coffea canephora*, known as robusta coffee, is used primarily to make instant coffee.

Tea (*Camellia sinensis*), also a small tree or shrub, is indigenous to India and China, where most of the world's production is concentrated today.

Figure 11–7 Cacao tree with pods. (Courtesy of Jorge Soria.)

The young leaves are carefully collected and then sorted to yield the various grades of tea. The postharvest processing is responsible for producing the different flavors of tea. Green teas are produced by drying and rolling the leaves, and black teas result from a fermentation during the drying process. It probably will come as no surprise to learn that Great Britain is the world's greatest importer of tea.

Chocolate and cocoa come from the cacao plant (*Theobroma cacao*—theobroma, from the Greek, means "food of the gods"), native to lowland tropical America and apparently first domesticated in Mexico. Details of its origin, like that of many of our domesticated plants, remain to be elucidated. When the Spanish reached Mexico, they found chocolate to be a prized drink among the Aztecs. Cacao beans, in fact, were once considered so valuable they served as currency. The cacao plant is a small tree with rather large leaves and is unusual in that it bears its small flowers, and eventually its pods, close to the trunk and branches. The pod, a specialized berry, yields several rather large seeds or "beans." Following fermentation (a rather odorous process), drying, and roasting, the seeds are ready to be ground. The whole bean gives us chocolate, which is a rich food as well as a tasty drink, for it contains about 30 to 50 percent oil, 15 percent starch, and 15 percent protein. Cocoa is produced by removing most of the fatty oils, which then are used as cocoa butter. Western Africa has replaced tropical America, where diseases have always plagued the trees, as the world's principal region of cacao production.

Two other genera, belonging to the same family as cacao, have seeds that contain caffeine. Seeds of the cola tree (*Cola acuminata* and *Cola nitida*), native to tropical Africa, have long been used in soft drinks. Seeds of guaraná (*Paullinia cupana*), a sprawling shrub or vine of the Amazon basin, are also used to prepare a soft drink that has sometimes been referred to as the national beverage of Brazil. In recent years it has become increasingly popular in other parts of the world, including the United States. Caffeine that is removed in making decaffeinated coffee is also used in some soft drinks.

Maté, or Paraguay tea, although less widely known than the previously discussed beverage plants, can hardly be called a minor beverage, since it is drunk by more than twenty million people in South America

Figure 11–8   Opened cacao pod, exposing the individual seeds, or beans. (Courtesy of USDA.)

and some is exported to Europe and North America. Maté comes from the leaves of *Ilex paraguariensis,* a relative of the holly tree, and is cultivated in Brazil, Paraguay, and Argentina. The processing of the leaves and the preparation of the drink are somewhat similar to the methods used for tea. Traditionally in South America maté is drunk from a gourd cup through a metal straw.

I have already mentioned alcoholic beverages in various places in this book, and it should be quite obvious by now that many different plants can be employed to prepare them. Few, however, can rival the grape. Grapes are, of course, one of our widely used fruits; they can be used fresh or as raisins, but most grapes go into the production of wine, with smaller amounts used to make brandy and cognac. The grape native to the Near East or surrounding area, *Vitis vinifera,* although not one of our earliest domesticates, is of considerable antiquity as a cultivated plant. Presumably the origin of wine, which is simply fermented grape juice, was not very complicated: someone squeezed some grape juice and let it stand, and wild yeasts converted some of the sugar in the grape to alcohol. Its preparation through the years, however, has become more elaborate.

The Bible tells us that Noah planted a vineyard, and in fact wine is mentioned no fewer than 165 times in the Bible. Wine had probably been around for some time before it reached the Greeks and Romans, who made considerable improvement in the "art" of wine making. The grape vine was carried to France in 600 B.C. and after Christianity became established monasteries played a significant role in establishing some of the great vineyards there. France and Italy were to become the world's foremost wine countries. Today Italy produces more wine than France, but France is generally regarded as producing the world's best wines. In the latter half of the eighteenth century there were devastating losses of the vines throughout Europe from diseases, and not until it was found that stems of *Vitis vinifera* could be grafted to rootstocks of the American species was there a recovery. It was perhaps only fitting that resistance to the diseases should have come from the American rootstocks, for they came to Europe with the introduction of American vines in the first place. This was also the time of Louis Pasteur's great discoveries, some of which contributed directly to an improvement of the wine industry.

Attempts were made to grow *Vitis vinifera* in eastern North America soon after it was settled, but both the humid climate and the cold winters were unfavorable, and it was to be some time before people turned to the American species and achieved success. In 1852 the Concord grape (named for the town in Massachusetts) came into being, either as

A

B

Figure 11–9 *A.* California vineyard in winter. (Courtesy of Wine Institute.) *B.* Harvesting the grapes. (Courtesy of Wine Institute.)

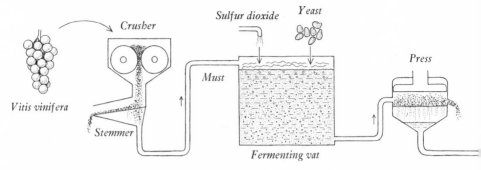

Figure 11–10 The method of producing wine in California is similar to that of Europe. Shown here are the steps in the production of red wine. The grapes are crushed to produce a "must." The must goes to a fermenting vat where yeasts transform the sugar into alcohol and then to a press for the removal of the skin and seeds. The wine then moves to settling

a mutant of the native fox grape or as a hybrid with *Vitis vinifera*. It was an immediate success, and its cultivation spread halfway across the continent in two years. The Concord and other native American grapes and hybrids of these with *Vitis vinifera* are the sources of the wines of the Great Lakes region today. The Old World species, however, was very successful in the Mediterranean climate of California, where it was introduced by the Spanish late in the eighteenth century and today is the basis of the California wines. New methods of preparing wine, including mechanical harvesting and aging in stainless steel tanks, were introduced in California and have spread to Europe. *Vitis vinifera* is also fairly widely grown for the production of wines in other parts of the world, including Argentina and Chile, where the climate is suitable.

In the United States per capita wine consumption has been increasing over the past several years and is now about two and a half gallons a year. This is quite a contrast to France where the annual per capita consumption is nearly twenty-one gallons, and in Italy and Portugal it is close to that amount. In all of these countries, however, consumption has decreased and is now about six gallons less than it was ten years ago.

Spices

In their search for edible plants, primitive people must have discovered many of those that now supply us with spices and condiments, and they probably learned to use these aromatic plants to make food more flavorful or to help cover up the taste and odor of food that had passed its prime. In time some of the spice plants became intentionally cultivated,

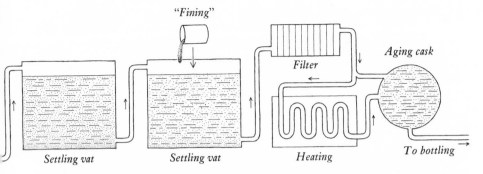

and with the Romans the spices came into their own in the art of cooking. Through Marco Polo's accounts Europeans became aware of the wealth of the spices of the Far East, and it was partly an attempt to secure these spices that led to the great ocean voyages of discovery in the fifteenth and sixteenth centuries.

Though hardly essential to nutrition, spices often made eating more enjoyable, and a large number of spice plants is found among our domesticated species. Reference has already been made to red pepper and mustard. The botanical families that have made the greatest number of contributions to our spices are the mint family, or Labiatae, and the parsley family, or Umbelliferae. The former has given us basil, marjoram, oregano, rosemary, sage, savory, and thyme, as well as spearment and peppermint; the latter has contributed anise, caraway, celery, chervil, coriander, cumin, dill, fennel, and, of course, parsley. The orchid family, which in number of species is the largest in the plant kingdom, has supplied only one plant that graces our food. Extract of vanilla comes from the pod of the vanilla orchid (*Vanilla planifolia*), native to tropical America and grown in many tropical areas, particularly Madagascar and Mexico. Although the plant is still cultivated for use as a flavoring, today most vanilla flavoring is made synthetically.

By far the world's most important spice is pepper (*Piper nigrum*), which at present accounts for one-fourth of the world's commerce in spices. The pepper plant is a member of the family Piperaceae and is not at all related to the *Capsicum* peppers previously mentioned. Apparently first domesticated in India, pepper became one of the first trade items between Europe and the Far East. Like many spices of the time, it was widely used in medicine as well as in seasoning and preserving food. The pepper plant is a woody vine, climbing to heights of thirty feet or more. The inconspicuous clusters of flowers each produce fifty to sixty fruits known as peppercorns. After drying, the whole peppercorns are ground to produce black pepper. If the hull is first removed, the ground product is white pepper. The plant is adapted to hot, wet, tropical regions and today the bulk of the world's supply is produced in India and Indonesia. The United States is the chief importer.

Another spice, nutmeg, received some notoriety as a hallucinogen a few years ago. Its use for this purpose is actually quite ancient, and it was recognized early that overindulgence could be lethal. Nutmeg comes from a tree, *Myristica fragrans*, native to the Malay Archipelago and now cultivated in the West Indies as well as in its homeland. The pulverized seeds produce nutmeg, and a second spice, mace, is prepared from the outer growth of the seed.

There are, of course, many other plants used for food. A few of the well known and widely used, as well as many lesser known, have not been included in this survey. The present account should, however, give some indication of the great diversity of plants that serve in our diet.

CHAPTER TWELVE

Multiplier of the Harvest

During this investigation we shall see that the principle of Selection is highly important. Although man does not cause variability and cannot even prevent it, he can select, preserve, and accumulate the variations given to him by the hand of nature almost in any way which he chooses; and thus he can certainly produce a great result.

CHARLES DARWIN, *The Variation of Animals and Plants under Domestication*

We are now entering an epoch of differential ecological, physiological, and genetic classification. It is an immense work. The ocean of knowledge is practically untouched by biologists. It requires the joint labours of many different specialists—physiologists, cytologists, geneticists, systematists, and biochemists. It requires the international spirit . . . We do not doubt that the new systematics will bring us to a new and better understanding of evolution, to a great increase in the possibilities of governing the processes of evolution, and to great improvement in our cultivated plants and domestic breeds of animals.

N. I. VAVILOV, *The New Systematics*

Domesticated organisms differ from their wild progenitors, often dramatically. Modern evolutionary theory provides an understanding of how these differences came about. The heredity of a plant or animal is controlled by its genes. Although genes are ordinarily very stable and copy themselves exactly generation after generation, spontaneous changes in them do occur. Such changes are known as gene mutations, and they may be passed on to the following generations. Through sexual reproduction genes are recombined in various ways, and thus it is possible for a species to "try out" various combinations of genes. Together mutation and recombination provide variation, the raw material of evolution. The force that acts upon this variation to produce change in a species is natural selection, which operates through differential reproduction. Thus, an organism with a particular combination of genes may produce more offspring than those with other combinations, or more of its offspring may survive and reproduce. In time such a successful combination will replace those that are less successful.

The same evolutionary factors—mutation, recombination, and selection—operate in the plants and animals that become closely associated with people. People also become a factor, however, in that human selection, in addition to natural selection, provides a guiding force. At first the selection by people was unconscious because no deliberate action was taken to improve the plant or animal, but it soon became intentional. The organisms became more and more dependent upon humans for their survival and often lost the ability to survive under natural conditions.

With plants that do not produce seeds by sexual reproduction and whose propagation is strictly vegetative—by tubers, roots, or other means—the evolution would be somewhat different in that no recombination of genes could occur. Once a vegetatively reproduced crop gives rise to a mutant type, all of its offspring would express the new character. Thus, conscious human selection may have operated from when these crops first came into cultivation. It should be pointed out that in some vegetatively propagated crops the plants still flower and may produce seeds by sexual reproduction. Consequently, there is a possibility of seedlings occurring that could show a recombination of characters. If some of these plants had desirable characteristics, they then could be perpetuated asexually.

Domesticated plants differ from their wild relatives in various ways: larger size, particularly of fruits or seeds; loss of natural means of dispersal; loss of delayed and irregular germination of seeds; simultaneous ripening; loss of toxic or bitter substances; loss of mechanical means of protection, such as spines; and changes in color of fruits or seeds. These changes resulted in a more desirable plant but one often ill adapted to survive in nature.

The loss of the natural means of dispersal will serve to illustrate how unconscious selection could have operated to produce a change. When most cereals were first brought into cultivation, the plants had a brittle fruiting stalk that would shatter readily, allowing some of the grains to fall to the ground before they could be harvested. A mutant type that had a nonbrittle fruiting stalk might appear occasionally and would hold all of its grain until harvested. Among seeds saved for planting there would likely be a high proportion from the plants with nonbrittle stalks. After the next year's sowing and harvest, still more seeds from the mutant strain would be collected. This process, if repeated year after year, would lead to more and more of the plants with nonbrittle stalks appearing in the following generations. For the nonbrittle type to replace the wild type completely, time would be required. The process would also depend to a large degree upon the nature of the gene or genes controlling this particular character. Conscious selection could have occurred, with a more rapid fixation of this character, had people realized the advantage of nonbrittle stalks and saved only seeds of this

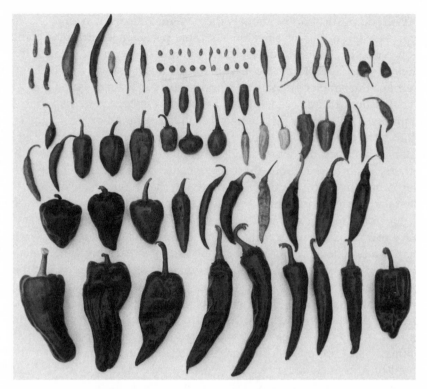

Figure 12–1 Variation under domestication in the chili pepper (*Capsicum annuum*). The wild plants have small, red, erect, deciduous, extremely pungent fruits (*upper row, center*). From these have been developed an array of forms, varying greatly in size, color, and pungency and including pendent and persistent fruits (*shown with stalk*) as well as erect and deciduous types.

type for planting, as has been done today with wild rice. It seems likely, however, that this character, as well as many of the others given above, changed through unconscious selection rather than through a deliberate act of the first seed planters.

How early intentional selection was practiced is not known, but there may have been an early selection for nonspiny types of plants and unusual seed or fruit colors. People might have attached significance to color, and, if the seeds were saved for planting to the exclusion of the normal seeds, there could have been a rapid change. People might have also attached special significance to large seeds, but if these were saved for planting it is unlikely that the large-seeded types would have been recovered in the next generation, for size is usually controlled by many genes. It is more likely that large seeds resulted from unconscious selection over a period of time.

With domestication in animals there were changes in size, color, and hair as well as in the skeleton and soft parts. In contrast to plants, however, the earliest domesticated animals were often smaller than their wild counterparts. The reason for this is not known with certainty. Perhaps the smaller animals were more easily managed. Possibly it was a matter of nutrition; wild animals perhaps were better fed than those in captivity. Quite often the changes resulted in the retention of juvenile characters to the adult state, as is evident in dogs. The domesticated animals also became more docile. This change, like many or all of the others, likely came about through unconscious selection; if the more fierce animals didn't escape from captivity, they would have been put to death and eaten. Intentional selection also probably occurred very early, if not from the beginning of domestication. For example, people must have realized quite early that only one or a few males were necessary to perpetuate a flock or herd. We might suppose that only certain males were kept for service and that these may have been chosen for their color, the shape of their horns or their lack of horns, their docility, and so on. Later, of course, there would have been selection for milk, wool, or meat production and the other characteristics that we regard as important today. Such selection would lead to a gradual change in the animals as in the plants. The length of time required for the establishment of a new character would depend upon the nature of its inheritance and the intensity with which breeding was managed.

How long did it take for a plant or animal to become domesticated? A precise answer to this question is impossible. Even if the archaeological record were complete, it would frequently be difficult to decide exactly when an organism qualifies as a domesticate, for the process of domestication is a gradual one. If, as supposed above, unconscious selection were important in the early stages of the evolution, we may also suppose that the process of domestication began when people started to keep animals and cultivate plants. The amount of time required would vary with the organism, its method of reproduction, the mutation rate, and the intensity of selection. Annual plants in which a new generation is grown every year could have changed remarkably in the space of a few generations. Tree crops with their longer generation time would have evolved much more slowly.

De Candolle pointed out a hundred years ago in his *Origin of Cultivated Plants* that no new basic food plants have been domesticated in historical times. This statement is still true and applies to animals as well as plants. How could primitive people have made such wise choices in their selection of plants and animals to domesticate? People must have had an intimate acquaintance with all the food resources in their environment when they began the process of domestication. We don't know if they started with the plants and animals that were difficult to obtain or with those that were common. It would seem reasonable that they started with those that they regarded as important—whether for religious reasons or because these were their favorite foods.

Among the most important early domesticated plants were annuals with large or numerous seeds. Many of them probably were originally weeds that grew in open or disturbed habitats. In a sense they were preadapted to cultivation, since they were prolific and could mature seeds in the space of a few months. It was therefore not entirely accidental that they gave rise to many of our most important food plants.

I pointed out earlier that certain animals—cattle, sheep, and goats, for example—were suited to human use since they could live on a diet that did not compete with our ancestors'. Moreover, the fact that they were gregarious animals rather than solitary meant that they could be more readily kept and managed. Some attempts were made to domesticate other animals as well. We know, for example, that the ancient Egyptians did so.

It is probable that many other plants and animals could have been domesticated. In fact only a fraction of 1 percent of the higher plants has been domesticated. Once people had their basic foods, apparently there was little incentive to domesticate others. It would have been unprofitable to spend the time to bring other species under their control for basic crops. The domestication of plants for special purposes, of course, did continue and is continuing today.

Domesticated plants and animals are extremely variable and usually comprise many races or varieties. The variability probably started quite early. When the domesticates spread to new areas either by the movement of people or by trade, they could have been subject to new types of selection, both natural and human. Thus variant types would have come into existence. Later, with an exchange between people, there would have been spontaneous hybridization between the newly different types of a particular plant or animal. Recombination in the offspring of the hybrids would provide new combinations of the parental characters. Natural hybridization between the domesticate and its wild progenitor or a closely related species also may have occurred; these hybrids would provide a new source of variability. If some of the variants had desirable characters, then those would have been perpetuated. It seems likely that hybridization followed by periods of isolation that allowed the establishment of new types provided the most rapid evolution. Some of the plants would have provided their own means of isolation by becoming self-pollinated. Many cultivated plants, including several of the cereals,

A

B

are self-pollinating, although the wild types that gave rise to them are habitually outcrossers.

For a long period of time conscious selection continued to be the main, if not the only, method for intentionally improving domesticated plants and animals. Spontaneous or natural hybridization undoubtedly contributed to some beneficial changes, however. Intentional hybridization of animals may be quite ancient, but hybridization as a deliberate way of improving plants is fairly recent. The first well-documented artificial plant hybrids were made in the eighteenth century, and the aim was not to improve the plants but to prove that they reproduced sexually. Gradually intentional hybridization became used to improve the plants. However, until the rediscovery of Mendel's laws in the first years of this century, the breeders were still largely working in the dark. Modern genetics gave plant and animal breeding a firm scientific basis, and planning replaced accident in breeding work.

Today selection is still an important part of any breeding program, but it is accomplished in far more sophisticated ways than in past centuries. Hybridization followed by backcrossing, both within and between species, continues to serve as the principal tool of the breeder. Inbreeding followed by hybridization led to the greatly increased yields of maize. Hybridization in wheat and rice has been responsible for the development of the new "miracle" seeds that produced the Green Revolution. Many other domesticated plants and animals have been greatly improved through intentional hybridization.

With the artificial induction of mutations by x-rays in the late 1920s by H. J. Muller came the realization that the breeder had a potential new tool. Instead of having to depend upon nature to supply all of the variation, breeders could induce change by using x-rays and other mutagenic agents. Seeds, seedlings, or whole plants were treated in the early days of mutation breeding. Recently the process has been speeded up in some plants by separating out single cells of a plant and growing these on cultural media. The spontaneous mutation rate in cell culture is often high. The cells may be treated with mutagenic agents as well. The mutant cells can then be grown into mature plants. Mutations occur more or less at random, and chance is involved in securing a desirable mutation. In fact, many, if not most, of the mutants secured are deleterious. Although mutational breeding has not lived up to our early hopes, a number of useful varieties has been created by this process.

Figure 12–2   A. Detail from Albrecht Dürer's *The Prodigal Son amid the Swine* (1496). *B.* Present-day sows with young. Breeders have considerably changed the appearance and productivity of the pig. (Courtesy of USDA.)

Figure 12–3 Rice breeder pollinating a flower to secure hybrid seed. (Courtesy of Rockefeller Foundation.)

Many of the major food plants—wheat, sugar cane, potatoes, sweet potatoes, and bananas among them—are polyploids. Just as scientists discovered how to induce mutation, they also found that polyploids could be created artificially. By treating plants with the chemical colchicine, for example, they can induce chromosome doubling, which gives rise to polyploid plants. Great hopes were once held for the production of economically important plants by use of this method, for polyploids are often larger and more hardy than their diploid progenitors. These hopes have not been entirely realized, but some valuable ornamentals, forage plants, and fruits, the most interesting perhaps being the seedless watermelon, have been developed. The understanding of polyploidy, however, has led to tremendous improvement in some of our polyploid crops. It has, for example, allowed desirable genes from the wild diploid species of goat grass (*Aegilops*) to be transferred to the bread wheats.

Efforts to secure haploids have also been successful in some crop plants and are proving of use in plant breeding. Haploid plants, like the pollen and the egg, have only one set of chromosomes and do not ordinarily produce seed. By doubling the chromosome number of such plants, a seed-producing plant may be obtained, however. Such a plant will be completely homozygous, each gene being represented by an exact counterpart, hence true breeding. These methods provide a way of securing "instant inbreds" when inbreds are ordinarily obtained only after many generations of self-pollination. Inbreds have been important in producing commercial hybrids in corn and several crop plants.

The opinion has been expressed that the traditional methods of breeding may have about reached their limits as far as improving the major crops such as wheat, corn, and rice. There is no agreement on this point, however, and certainly there are many crops in which a great deal is yet to be accomplished with the old methods. Nevertheless, it is evident that new approaches are called for, and fortunately some are now available. The developing field of molecular biology holds great promise for improving domesticates by genetic engineering. With traditional methods of breeding only members of a species or closely related species belonging to the same genus can be hybridized in order to transfer useful genes to the domesticated organism. With molecular techniques it has become possible to transfer genetic material directly between unrelated species. Thus far no great achievements have appeared, but several of some significance appear on the horizon. Crop plants that are virus resistant, insect and herbicide tolerant and suffer reduced damage from frost are now in advanced experimental states. Some people fear that genetic-engineered organisms might escape and damage the environment. This danger has probably been exaggerated for higher plants. After all, in a sense, people have been carrying on a form of genetic engineering ever since they started domesticating plants and animals several thousand years ago. True, at times some of the plants have escaped and become weeds—as sorghum has—and animals have become feral—goats, for example—and caused considerable damage. The same could happen to the current creations. Clearly, careful testing and controls are called for. Finally, it should be pointed out that molecular breeding will not replace the traditional methods but join with them in improving the domesticated organisms.

For breeding work to continue to be successful it is essential that genetic diversity be available. At one time the farmers' fields provided a valuable reservoir of genes. But with the wide-scale adoption of hybrids and new highly productive varieties around the world, fewer and fewer varieties are grown, resulting in a serious reduction in genetic diversity. The Green Revolution has contributed to this phenomenon, for as the new varieties of wheat and rice are adopted and widely accepted, the old ones disappear. Although the older varieties may be much less productive than the newer ones, they are not worthless, for they may contain genes that could be valuable in the future. The danger of a single variety being widely grown was well illustrated when a blight disease struck the corn crop in the United States in 1970. An earlier example was the potato blight in Ireland (see Chapter 8). Had there been greater diversity in the potatoes, there likely would have been far less damage to the crop. The less diversity there is in the crop the greater its vulnerability.

At the same time it was recognized that genetic erosion was occurring in our domesticates, it became apparent that their wild relatives could also be lost and with them another important source of potentially useful genes. With the current destruction of habitats around the world, wild species are being exterminated at an alarming rate. The new perennial teosinte of Mexico that may have much to offer for the improvement of maize exists today in an area of only six acres (see Chapter 5). Even if the wild species are not lost, their genetic variability could become very restricted as their range contracts. The noted biologists and conservationists Paul Ehrlich and Anne Ehrlich have written that the loss of genetic variability may be the most serious problem facing humankind.

One way to preserve genetic diversity is to establish gene banks. Although a few gene banks have existed for many years, the recent recognition of the great importance of genetic diversity has led to the establishment of more. The effort was promoted by the founding of the International Board for Genetic Resources in 1974. Fifty gene banks were in operation in various parts of the world by 1986. Seeds of crop plants, and sometimes their wild relatives, are stored there under low temperatures. Although low temperatures greatly extend the life span of most seeds, plants still have to be grown periodically to renew the seeds. Crops that do not normally produce seeds are maintained in nurseries; and with the advances in cellular biology it is becoming possible to store small amounts of tissues in test tubes on nutrient agar, which has great advantage over the use of nurseries. For the gene banks to be an effective source of genes for the breeder, the material in them has to be catalogued and evaluated. Gene banks are not inexpensive operations, and for the nations of the Third World to maintain them often requires outside financial assistance. The world is now in a much better position as far as genetic resources are concerned than it was only a few years ago, but much more collection of materials, storage, and evaluation remains to be done.

An example of the importance of gene banks is provided by the U.S. Tomato Genetic Center at the University of California at Davis. Its bank has twenty-six hundred accessions of the tomato, including many collections of wild species obtained by the center's director, Charles Rick, in his numerous trips to South America. These have proven of immense benefit in the improvement of the tomato. For example, resistance to thirty-two diseases has been discovered, and resistance to sixteen of them has already been bred into commercial varieties.

In the past there has been, with a few exceptions, a free interchange of seeds among breeders and investigators in different countries. The recent attention given to genetic resources has raised questions concerning their ownership. Are genetic resources a common global heritage or do they belong to the nations in which they are found? Are there exceptions, such as the elite lines developed by the breeder? At a conference of the United Nations Food and Agriculture Organization (FAO) in 1983 a resolution was adopted known as the "Undertaking" that called for unrestricted access to plant genetic resources, including special genetic stocks such as elite and current breeders' lines. Eighty-one countries had accepted the resolution by 1987, but some have balked, particularly those nations with well-developed seed industries. Some of the seed companies in the developed nations feel strongly that, if they have developed an elite line of plants at a considerable financial investment, they deserve some compensation for the release of these stocks. The developed nations have been characterized as rich in technology but poor in genes, whereas the opposite is true of the undeveloped nations. This statement is an oversimplification but does hold some truth. Certainly the United States is gene poor, for only one major crop, the sunflower, has been developed within its borders. Some people in the Third World feel that industrial nations control the genetic resources and that these will be exploited by multinational corporations. Thus, what the press has called "seed wars" are now being waged. No solution completely acceptable to all parties has emerged, and it is imperative that a system for exchanging germ plasm be found that will benefit all of humanity.

For future advances to be made in the breeding of plants and animals we shall continue to need a strong investment in basic research, much of which in the past has come from the universities. Two kinds of research may be distinguished—pure, or basic, and applied. Direct attempts to improve our food plants and animals, of course, would be applied research. The techniques that have made them possible come from basic research. Gregor Mendel did not set out to make a better pea when he began his hybridization of various kinds of peas. He was simply curious, and out of his curiosity came the laws of genetics, the basis of plant and animal breeding. A more recent example is the research of James D. Watson and Francis Crick. Their curiosity concerned the nature of the gene and as a result sophisticated genetic engineeing has now become feasible. Clearly the breeders and other biologists can contribute further to the elimination of hunger and malnutrition in the world, but they alone cannot be expected to solve the problem.

CHAPTER THIRTEEN

Let Them Eat Cake?

> And he gave it for his opinion, that whoever could make two ears of corn, or two blades of grass, to grow upon a spot of ground where only one grew before, would deserve better of mankind, and do more essential service to his country, than the whole race of politicians put together.
>
> JONATHAN SWIFT, *Gulliver's Travels*

> The time has come to make peace with each other so that we can make peace with the earth.
>
> LESTER R. BROWN AND EDWARD C. WOLF, *The State of the World*

> When there are starving people in the world, it seems wrong that so many of us Americans eat as much for entertainment as for nourishment.
>
> ANDY ROONEY

The population of the world is now over five billion. How many of these people are hungry is not known, but recent estimates show four hundred to seven hundred million people being undernourished and thirteen to eighteen million dying each year from starvation and hunger-related diseases. Nearly all of the malnutrition occurs in the Third World—what some people call the developing nations or simply the poor nations. Most of these are in the tropics—Africa, Latin America, and Asia—where most of the world's people live.

It is generally agreed that the world now produces enough food so that no one needs to be malnourished. The primary reason for malnutrition is poverty. People with money do not go hungry even in Bangladesh or Haiti, two of the countries with grave problems. The reasons for poverty are primarily political, economic, and social, not biological, and hence fall beyond the scope of this book; therefore, the following remarks will be confined largely to biology.

Although there is currently enough food to provide all people with an adequate diet, there could well be serious food supply problems in the future. The population is increasing at a rate of 1.7 percent a year; it is projected to be over six billion before the end of this century and to double sometime early in the next century. Unless something is done soon, the number of hungry people in the world will have increased greatly as we enter the twenty-first century.

Can we increase food production and decrease population growth? There have been some encouraging signs. Food production expanded at an annual rate of 3 percent between 1950 and 1973 and has increased around 2 percent each year since then. This trend resulted from cultivating more land, much of it by irrigation, using the new varieties of wheat and rice, and expending greater energy. The world's most populous nations, China and India, have become self-sufficient in food, and Indonesia has begun to export rice. Some Third World nations have advanced toward controlling the growth of their populations, although many of them still show a growth rate of 3 percent or more. India and perhaps China will again become food importers as their numbers continue to grow. The United States, previously a strong supporter of family planning, withdrew its financial support of UN population activities under the Reagan administration.

There are a number of problems that do involve biology—particularly ecology—that must be solved if food production is to meet the needs of the growing number of people. In the *State of the World 1986* Lester R. Brown, one of the world's foremost authorities on agriculture and the environment, lists the resource depletions that are having adverse effects on the world's economy (Table 13–1). Each of these resources has a bearing in one way or another on food production.

1. Not too many years ago we were looking to the oceans to provide greater amounts of food. Recently, however, the yields have been increasing at the rate of only 1 percent a year. Both overfishing and pollution have been responsible for this failure. Saving the whales has received wide attention, but we must be concerned with saving the other organisms in the ocean as well. And fresh waters are in trouble too. Many lakes are now "biologically dead" because of pollution.

2. According to the World Wildlife Fund, the tropical rain forests are being destroyed at the rate of fifty acres a minute. Although the destruction of the world's forests, particularly in the tropics, has received wide attention in recent years, efforts at slowing it have only just begun. The forests are cut for lumber, for firewood, now in short supply in many parts of the world, and to provide land for crops or pasture. The clearing of the forests has serious ecological consequences. It is causing species extinction not only of trees but of other plants and animals that live in these ecosystems as well. The trees absorb large amounts of carbon dioxide and their loss is accelerating the greenhouse effect. Moreover, the cleared lands are subject to severe water runoff and with it the

Table 13–1 Resource depletions adversely affecting global economy

Resource	Extent of depletion
Fisheries	Rapid growth in world fish catch during fifties and sixties now history; overfishing often the rule, not the exception. Fish catch per person, including fish farming, down 15 percent since 1970. Biggest consumption cuts are in Third World countries such as the Philippines.
Forests	World's tropical forests disappearing at 2 percent per year. Far faster in West Africa and southeast Asia, where moist tropical forests will have virtually disappeared by end of century. Previously stable forests in temperate zone now suffering from air pollution and acid rain. Dead and dying forests plainly visible in West Germany, Czechoslovakia, and Poland.
Grasslands	Excessive pressure on grasslands, closely paralleling growing pressure on forests and soils, has led to deterioration, which is most advanced in Africa and Middle East. Herd liquidation in pastoral economies of Africa now commonplace.
Oil	Increases in the price of oil, the principal commercial fuel, has sharply reduced world economic growth since 1973. Part of decline is due to ill-conceived responses to oil price hikes, notably the heavy borrowing by Third World countries. Progress in developing renewable alternatives is lagging.
Soil	Soil erosion exceeding new soil formation on 35 percent of world's cropland. World losing an estimated 7 percent of topsoil per decade. Effects most evident in Africa, where 40 percent of people live in countries where land productivity is lower than it was a generation ago.
Water	Growing water demand exceeding sustainable supplies in many locations, leading to scarcity. Falling water tables now found on every continent and in key food-producing regions. In some areas, including portions of the United States, water being shifted out of irrigated agriculture to satisfy growing residential demands.

Source: Lester R. Brown, *State of the World, 1986* (New York, Norton). Used by permission.

erosion of soil. The wide-scale flooding in Bangladesh in 1988, which brought major suffering to the people, was largely the result of forest clearing to the north. Reforestation projects are active in many parts of the world, but it will take a long time for them to have much effect.

3. If we eat less meat in the future, we should give up the animals that

are fed on grain, not those raised on rangeland, for there are many areas of the world where the land can support grazing but is ill adapted to growing crops. Animals in these areas can convert nonhuman food into protein for human use. Both overgrazing, because of the greater number of animals, and climatic factors, however, bode ill for increasing the amount of such protein in the future. Although most people in the Third World depend mainly on plants for their protein, there are some who depend on meat and milk or on the sale of animal products to obtain cash to buy other food.

4. Much of today's agriculture requires large amounts of energy, usually in the form of petroleum. We depend, for example, on fuel to run tractors, pump water, and bring food to market. At one time it was thought that nuclear power would help solve the energy problem, but with growing concern over the risks involved, that is less likely. Geothermal sources, wind, water, and biomass may replace petroleum instead. Biomass could already be used much more than it presently is. Crop plants, such as sugar cane, can be used for alcohol production, another source of energy, but that would mean a sacrifice of land that could be used for food production. Moreover, our use of energy can be made more efficient. The staggering foreign debts of many nations stem in large part from importing fuel. The price of oil reached its peak in 1981. With the present glut of oil in the market, prices are lower but there is no assurance that these prices will remain stable for long. Despite the discovery of huge deposits of petroleum in Mexico, that country has a huge foreign debt. The debts of Third World nations make it difficult for them to raise the capital needed to improve food production.

In addition to fuel energy is needed for the manufacture and application of fertilizers, herbicides, and pesticides. The increase of their use since 1950 has been another factor in the increase of food production. Herbicides not only kill weeds but reduce soil erosion by eliminating the need to plow and cultivate cropland. All of these chemicals, however, pollute the environment. Organic farming in which no chemicals are used is now employed, but only on a very small scale. Although widespread organic farming seems unlikely, at least for a long time to come, greater effort must be made to reduce the use of toxic chemicals in agriculture. Biological control of pests, that is, using their natural enemies, can be more widely employed. This technique is hardly new, for it was used as early as 1888 to control scale insects on the citrus crop in California. It is now available for a number of other crops as well.

5. Soil erosion is hardly a new problem, but it remains a critical one. For example, it has been reported that in recent years over one billion tons of top soil have been lost annually to erosion in highland Ethiopia.

Increased food production in the last thirty years has come about in part by cultivating new land. Much of this land is marginal for agriculture and often subject to severe erosion. Of course, serious erosion occurs in many places, including the United States, on the lands long used for agriculture. In addition to land becoming unfit for productive agriculture because of erosion, more and more land goes to urban development every year, and this will continue as the population continues to grow. If very little land is to be available for the expansion of agriculture in the future, perhaps some nonessential crops such as tobacco will have to be abandoned. With little prospect of increasing crop acreage, it becomes more obvious how important it is to improve productivity.

6. The amount of land irrigated for agriculture has nearly tripled in the last forty years. At the same time, some irrigated land has been lost because of salinization and waterlogging. The lack of fresh water appears to be the limiting factor in bringing more arid land under irrigation. Some techniques for more efficient irrigation can be more widely applied and it is possible that some crops can be bred to survive naturally in the dry areas. Nonetheless, the prospect of such areas producing much more food, with or without irrigation, does not appear too likely.

In addition to the factors discussed above global warming will have profound effects on agriculture. In fact, some people think that it has already begun. The cause of this phenomenon is the build up of various "greenhouse" gases in the atmosphere—chlorofluorocarbons, which are also destroying the stratosphere's ozone layer, nitrogen oxides, methane, and carbon dioxide—all of which are almost entirely the result of human activities. The primary problem is carbon dioxide which comes from the burning of fossil fuels mainly in the United States and the Soviet Union. The destruction of forests has also contributed to this problem because they have been the major means of absorbing excess carbon dioxide. As a result of the greenhouse effect it is predicted that in the not too distant future there will be higher temperatures, drier soils, and reduced water supply in some parts of the world. This would mean a drastic shift of vegetation zones and with it the present patterns of agriculture.

Although exact predictions cannot be made, it is likely that we will no longer be able to grow many of the major crops where they are grown today. At the same time, however, it is also likely that we will be able to grow crops such as corn and soybeans to the north of where they are today. In all probability the United States would no longer be a major exporter of food, although it should still be able to produce enough to feed itself. Global warming will also cause the polar ice caps to melt.

Figure 13–1 Carcass of a cow in drought-stricken area of Mauritania. (Courtesy of FAO.)

With the rising sea levels delta areas now devoted to agriculture, as well as coastal cities, will disappear. With the climatic change we can also expect a further extinction of both plants and animals. New varieties of crops, and perhaps some altogether new crops, can be developed to grow in hotter, drier regions, but this will take time and require a massive effort on the part of the breeders. How rapidly the global climatic change will take place is still uncertain, but there can be little doubt that it will have severe effects on food production.

Is there room for optimism? Some people feel that there is. They point to the recent increases in food production and assume that the trend will continue. Others seem to think that genetic engineering and other forms of biotechnology can solve all of the problems. Then, of course, there are the prophets of doom who think that it is already too late and that we face mass starvation in the next century, if a nuclear holocaust doesn't destroy us first. But if we recall William and Paul Paddock's 1967 book *Famine 1975! America's Decision: Who Will Survive,* we see that the authors were wrong, and we can only hope that the modern doomsday prophets will be as well. Many people who have investigated the subject feel that there may still be time to solve the world's ecological problems, to reduce the birth rates, and to increase food production, but that if we haven't made significant advances by the year 2000 it may be too late. However, given even the most optimistic predictions, hunger will still be with us in the future, and it will probably involve far more people than it does today. To attempt to solve the world's problems and to eliminate malnutrition the nations of the world need wise, able, foresighted, and honest leaders who have an understanding of ecological principles. Such leaders are now in short supply. And the problems aren't the leaders' alone. All of us must be involved.

Bibliography
Index

Bibliography

General

Bailey, L. H., and E. Z. Bailey. 1976. *Hortus third: A concise dictionary of the plants cultivated in the United States and Canada.* New York: Macmillan.

Barigozzi, C., ed. 1986. *The origin and domestication of cultivated plants.* Amsterdam: Elsevier.

Crosby, Alfred W., Jr. 1972. *The Columbian exchange: Biological and cultural consequences of 1492.* Westport, Conn.: Greenwood Press.

Economic Botany, the Journal of the Society for Economic Botany, published by the New York Botanical Garden, Bronx, New York. (A journal of applied botany and plant utilization.)

Farb, P., and G. Armelagos. 1980. *Consuming passions: The anthropology of eating.* Boston: Houghton Mifflin.

Harlan, J. R. 1975. *Crops and man.* Madison, Wisc.: American Society of Agronomy. (A textbook dealing with the origin and evolution of crop plants by an authority in the field.)

Hartman, H. T., A. M. Kofranek, V. E. Rubutzky, and W. J. Flocken. 1982. *Plant sciences: Growth, development and utilization of cultivated plants.* 2nd ed. Englewood Cliffs, N.J.: Prentice Hall.

Janick, Jules, Robert W. Schery, Frank W. Woods, and Vernon W. Ruttan. 1981. *Plant science: An introduction to world crops.* 3rd ed. New York: W. H. Freeman. (An introductory text covering botanical, technological, and economic aspects of agriculture.)

Kahn, E. J., Jr. 1985. *The staffs of life.* Boston: Little, Brown. (Chapters on maize, potatoes, wheat, rice, and soybeans.)

Klein, R. M. 1986. *The green world: An introduction to plants and people.* 2nd ed. New York: Harper & Row.

McGee, H. 1984. *On food and cooking: The science and lore of the kitchen.* New York: Scribner's. (Treats nearly all foods, their basic chemistry, nutrition, and much more.)

Purseglove, J. W. 1968–1972. *Tropical crops: Dicotyledons*, 2 vols., and *Tropical crops: Monocotyledons*, 2 vols. New York: Wiley. (A treatment of virtually all tropical economic plants, many of which are also cultivated in the temperate zones. In addition to detailed descriptions and many illustrations, there are remarks on pollination, germination, propagation, chemical composition, husbandry, pests and diseases, breeding, and origins.)

Reed, C. A., ed. 1977. *Origins of agriculture.* The Hague: Mouton. (Provides a variety of viewpoints concerning the origins of both plant and animal domestication as well as consideration of the early history of agriculture.)

Simmonds, N. W., ed. 1976. *Evolution of crop plants.* London: Longman. (An indispensable work for anyone interested in the origin and evolution of domesticated plants; cytotaxonomy, history, and prospects for all major crops, with notes on nearly all of the minor crops.)

Simpson, B. B., and M. Conner-Oyorzaly. 1986. *Economic botany: Plants in our world.* New York: McGraw-Hill. (Treatment of all important economic plants; now the standard textbook in the field.)

Tippo, Oswald, and W. L. Stern. 1977. *Humanistic botany.* New York: W. W. Norton. (An elementary botany textbook with emphasis on economic plants. Excellent reading for anyone wanting to know more about plants.)

Visser, M. 1987. *Much depends on dinner: The extraordinary history and mythology, allure and obsessions, perils and taboos, of an ordinary meal.* New York: Grove. (Includes chapters on corn, chicken, rice, lettuce, olive oil, lemon, and other subjects.)

Chapters One and Two

Anderson, E. 1952. *Plants, man and life.* Boston: Little, Brown, (Written for the interested layman; a rather unconventional but fascinating introduction to weeds, cultivated plants, and the botanists who study them. Some of the material regarding the origin of certain crops has been superseded by recent discoveries.)

Cohen, M. N. 1977. *The food crisis in prehistory: Overpopulation and the origins of agriculture.* New Haven: Yale University Press.

Flannery, K. V. 1986. *Guilá Naguitz: Archaic foraging and early agriculture in Oaxaca, Mexico.* Orlando, Fla.: Academic Press.

Ford, Richard I., ed. 1985. *Prehistoric food production in North America.* Museum of Anthropology, no. 75. Ann Arbor: University of Michigan Press.

Gaster, T. H., ed. 1964. *Sir James Frazer's the new golden bough.* New York: Mentor. (An abridged and somewhat revised edition of a classic work. Still probably the best source on myth, magic, and religion in relation to primitive and ancient agriculture.)

Harris, D. R., and G. C. Hillman, eds. 1989. *Foraging and farming: The evolution of plant exploitation.* Oxford: Oxbow.

Heiser, C. B. 1985. *Of plants and people.* Norman: University of Oklahoma Press.

Isaac, Erich, 1970. *Geography of domestication.* Englewood Cliffs, N.J.: Prentice Hall.

Jensen, A. E. 1963. *Myth and cult among primitive peoples.* Chicago: University of Chicago Press.

Rindos, D. 1984. *The origins of agriculture: An evolutionary perspective.* Orlando, Fla.: Academic Press.

Sauer, C. O. 1969. *Seeds, spades, hearths and herds: The domestication of animals and foodstuffs.* 2nd ed. Cambridge, Mass.: MIT Press. (Much of this book is a reprint of the original edition published in 1952 under the title *Agricultural Origins and Dispersals* and hence does not take into account new information now available. Nevertheless, it remains one of the most stimulating discussions of the origin of agriculture.)

Zohary, D., and M. Hopf. 1988. *Domestication of plants in the Old World: The origin and spread of cultivated plants in West Asia, Europe and the Nile Valley.* Oxford: Clarendon Press.

Chapter Three

Eaton, S., M. Shostak, and M. Konner. 1988. *The Paleolithic prescription: A program of diet & exercise and a design for living.* New York: Harper & Row.

Krause, M. V., and L. K. Mahan. 1984. *Food, nutrition and diet therapy.* 7th ed. Philadelphia: W. B. Saunders.

U.S. Department of Agriculture. 1985. *Nutritive value of foods.* Washington, D.C. (Gives nutritive values of more than nine hundred common foods.)

Whitney, E. N., and E. M. N. Hamilton. 1987. *Understanding nutrition.* 4th ed. St. Paul: West.

Chapter Four

Clutton-Brock, J. 1981. *Domesticated animals from early times.* Austin: University of Texas Press.

Clutton-Brock, J., ed. 1988. *The walking larder: Patterns of domestication, pastoralism and predation.* Oxford: Oxbow.

Cole, H. H., and W. N. Garrett. 1980. *Animal agriculture: The biology, husbandry and use of animals.* 2nd ed. New York: W. H. Freeman.

Harris, Marvin. 1985. *Good to eat: Riddles of food and culture.* New York: Simon and Schuster.

Mason, I. L., ed. 1984. *Evolution of domesticated animals.* London: Longman.

Zeuner, F. E. 1963. *A history of domesticated animals.* London: Hutchinson.

Chapter Five

Grist, D. H. 1986. *Rice.* 6th ed. London: Longman.

Iltis, H. H. 1987. Maize evolution and agricultural origins. In *Grass systematics and evolution,* eds. T. R. Soderstrom, K. W. Hilu, C. S. Campbell, and M. E. Barkworth, pp. 195–213. Washington, D. C.: Smithsonian Institution Press.

Kerby, K., and J. Kuspira. 1987. The phylogeny of the polyploid wheats *Triticum aestivum* (bread wheat) and *Triticum turgidum* (macaroni wheat). *Genome* 29: 722–737.

Oka, H. L. 1988. *Origin of cultivated rice.* Tokyo: Japan Scientific Societies Press.

Sprague, G. F., and J. W. Dudley, eds. 1988. *Corn and corn improvement.* 3rd. ed. Madison, Wisc.: American Society of Agronomy.

Wallace, H. A., and W. L. Brown. 1988. *Corn and its early fathers.* 2nd ed. Ames: Iowa State University Press.

Chapter Six

Barnes, A. C. 1974. *The sugar cane.* 2nd ed. New York: Wiley.

Chapter Seven

Duke, J. A. 1981. *Handbook of legumes of world economic importance.* New York: Plenum Press.

Weiss, E. A. 1983. *Oilseed crops.* London: Longman.

Chapter Eight

Cock, J. H. 1985. *Cassava, new potential for a neglected crop.* Boulder, Colo.: Westview Press.

Coursey, D. G. 1975. The origins and domestications of yams in Africa. In *Origins of African plant domestication,* ed. J. R. Harlan, pp. 383–408. The Hague: Mouton.

Jennings, D. L., and C. H. Hershey. 1985. Cassava breeding: A decade of progress from international programs. In *Progress in plant breeding,* I, ed. G. L. Russell, pp. 89–116. London: Butterworths.

Salaman, R. N. 1985. *The history and social influence of the potato.* Cambridge: Cambridge University Press. (Revised, with introduction by J. G. Hawkes.)

Simmonds, N. W. 1982. *Bananas.* 2nd ed. New York: Wiley.

Wang, J., ed. 1983. *Taro: A review of* Colocasia esculenta *and its potential.* Honolulu: University of Hawaii Press.

Yen, D. E. 1974. *The sweet potato and Oceania.* Honolulu: B. P. Bishop Museum.

Chapter Nine

Child, Reginald. 1974. *Coconuts.* 2nd ed. London: Longman.

Woodruff, J. G. 1970. *Coconuts: Production, processing, products.* Westport, Conn.: AVI.

Chapter Ten

Carter, J. F., ed. 1978. *Sunflower science and technology.* Madison, Wisc.: American Society of Agronomy.

Heiser, C. B. 1976. *The sunflower.* Norman: University of Oklahoma Press.

Chapter Eleven

Adams, L. D. 1985. *The wines of America.* 3rd ed. New York: McGraw-Hill.

Amerine, M. A., H. W. Berg, R. E. Kunker, C. S. Origh, V. L. Singleton, and A. D. Webb. 1980. *The technology of wine making.* Westport, Conn.: AVI.

Janick, J. 1986. *Horticultural science.* 3rd ed. New York: W. H. Freeman.

Purseglove, J. W., E. G. Brown, C. L. Green, and S. R. J. Robbins. 1981. *Spices,* 2 vols. London: Longman.

Rosengarten, F. 1984. *The book of edible nuts.* New York: Walker.

Wood, G. A. R. 1975. *Cacao.* 3rd ed. London: Longman.

Chapter Twelve

Diversity. A news journal for the plant genetics resources community. Washington, D.C.

Fraley, R. T., S. G. Rogers, R. B. Horsch, G. M. Kishore, R. N. Beachy, N. T. Tumer, D. A. Fiskhoff, X. Delannay, H. J. Klee, and D. M. Shah. 1988. Genetic engineering for crop improvement. In *Chromosome structure and function: Impact of new concepts,* eds. J. P. Gustafson and R. Appels. New York: Plenum.

Hawkes, J. G. 1983. *The diversity of crop plants.* Cambridge, Mass.: Harvard University Press.

Heiser, C. B. 1988. Aspects of unconscious selection and the evolution of domesticated plants. *Euphytica* 37:77–81.

Juma, Calestous. 1989. *The gene hunters: Biotechnology and the scramble for seeds.* Princeton, N.J.: Princeton University Press.

Kloppenburg, J. R., ed. 1988. *Seeds and sovereignty: The use and control of plant genetic resources.* Durham, N.C.: Duke University Press.

Plucknett, D. L., N. J. H. Smith, J. T. Williams, and N. M. Anishetty. 1987. *Gene banks and the world's foods.* Princeton, N.J.: Princeton University Press.

Simmonds, N. W. 1979. *Principles of crop improvement.* London: Longman.

Chapter Thirteen

Bennett, J. (with S. George). 1987. *The hunger machine.* Cambridge: Polity Press.

Bernardi, G. M., ed. 1985. *World food, population and development.* Totowa, N.J.: Rowman and Allanheld.

Brown, L. R., W. U. Chandler, A. Durning, C. Flavin, J. Hesse, J. Jacobson, S. Postel, C. P. Shea, L. Starke, and E. C. Wolf. 1988. *State of the world, 1988.* New York: Norton. (See also *State of the World* for the years 1984–1987, my chief sources for this chapter.)

Doyle, J. 1985. *Altered harvest: Agriculture, genetics and the fate of the world's food supply.* New York: Viking Press.

George, S. 1988. *A fate worse than debt.* London: Penguin.

Lappé, F. M., and J. Collins. 1982. *World hunger: Ten myths.* San Francisco: Institute for Food and Development Policy.

Longacre, Doris. 1976. *More-with-less cookbook.* Scottdale, Pa.: Herald Press.

Pimental, D. F., and C. W. Hall, eds. 1989. *Food and natural resources.* San Diego: Academic Press.

Richards, P. 1986. *Coping with hunger: Hazard and experiment in an African rice-farming system.* London: Allen and Unwin.

Wolf, E. C. 1986. *Beyond the green revolution: New approaches for Third World agriculture.* Worldwatch Paper 73. Washington, D.C.: Worldwatch Institute.

Yeşilada, B. A., C. D. Brockett, and B. Drury, eds. 1989. *Agrarian reform in reverse: The food crisis in the Third World.* Boulder, Colo.: Westview Press.

Index

Abaca, 158
Abelmoschus esculentus, 171
Abrin, 122
Abrus precatorius, 122
Acer saccharum, 111
Achard, 115
Acorn, 187
Actinidea chinensis, 187
Adonis, 25
Aegilops squarrosa, 70, 73
Agriculture: places of origin, 6–10; religious origin, 18–24; single or multiple origins, 10; theories of origin, 14–26; women's role in origin, 14
Aguardiente, 90, 114
Aje, 142
Akee, 151
Albizia, 121
Alcoholic beverages, 64, 86, 90, 114, 164, 192–194
Alfalfa, 121
Allium, 183
Almond, 187–188
Alpaca, 58
Amaranth, 108
Amaranthus, 108
Amino acids, 30–31, 104
Anabaena, 86
Ananas comosus, 186
Anderson, Edgar, 15
Angiosperms, 61
Angora, 43
Animals, 34–60; domesticated, definition of, 34; domestication of, 21, 31, 201; in preparing fields, 12; use of, 36
Ants, 15
Aphrodisiac, 136, 141
Apichu, 142
Apium graveolens, 183
Araceae, 149
Arachis hypogaea, 127
Aroid family, 149
Arracacha, 183
Arracacia xanthorrhiza, 183
Arrack, 164

Artiodactyla, 36
Artocarpus: altilis, 151; *heterophyllus*, 152
Ass, 52
Astragalus, 123
Aurochs, 43
Avena sativa, 106
Avocado, 183
Azolla, 86

Bacteria, 118
Bakanae, 87
Bamboo, 65–66
Banana, 153–158
Barley, 106–107
Batata, 138, 142
Beadle, George, 96, 97
Beal, William James, 99
Beans: adzuki, 126; asparagus, 124; black-eyed, 124; black gram, 126; broad, 122; butter, 124; common, 124–126; English, 122; fava, 122; golden gram, 126; green, 124; jequirity, 122; kidney, 124; lima, 124; lupini, 123; mat, 126; moth, 126; mung, 124; rice, 126; scarlet runner, 124; sieva, 124; soy, 129–133; tepary, 124; urd, 126; wax, 124; winged, 120; yam, 120; yard-long, 124
Beefalo, 48–49
Beer, 64, 68
Beet, 115
Beriberi, 86
Berries, 184
Beta vulgaris, 115
Beverages, 188–194. *See also* alcoholic beverages
Bezoars, 41
Bison, 48
Bligh, William, 151
Blighia sapida, 151
Blight: corn, 102, 205; potato, 137, 205
Boll weevil, 175
Borlaug, Norman, 77–78
Bos: indicus, 44; *taurus*, 43
Bounty, H.M.S., 151
Bourbon, 90

223

Braidwood, R. J., 6, 15
Brassica: campestris, 178; *napus,* 178; *oleracea,* 178
Brazil nut, 188
Bread, 68, 76
Breadfruit, 151–153
Breadnut, 152
Breeding, 203–207
Broccoli, 178
Bromelain, 187
Broomcorn, 106
Brown, Lester, 210
Brussels sprouts, 178
Bubalus bubalus, 49
Buckwheat, 108
Buffalo: American, 48; water, 49–51

Cabbage, 178
Cacahuate, 127
Cacao, 190–191
Caffeine, 188, 191
Callejon de Huaylas, 9
Calorie, defined, 29
Camellia sinensis, 189
Camote, 142
Canis familiaris, 37
Canola, 178
Capra hircus, 41
Capsicum, 182, 199; *annuum,* 182
Carbohydrates, 29
Carica papaya, 187
Carina moschata, 58
Carnauba wax, 159
Carob, 120
Carrot, 183
Carter, George F., 57
Carthamus tinctorius, 171–172
Carver, George Washington, 129
Cashew, 188
Cassava, 143
Castration, 44, 55
Catal Hüyük, 21
Cattalo, 48
Cattle, 21, 23, 43–48; in Africa, 45, 47; in India, 45–47
Cauliflower, 178
Cavendish banana, 156
Cavia porcellus, 58
Cavy, 58
Celery, 183
Cellulose, 29
Ceratonia siliqua, 120
Cereals, 61–110
Ceres, 64
Chenopod, 108

Chenopodium, 108, 110, 116
Chestnut, 187
Chicha, 90
Chicken, 55–57
Chickpea, 124
Child, Reginald, 166–167
Childe, V. Gordon, 15
Chocho, 123
Chocolate, 190
Cholesterol, 30
Christian, Fletcher, 151
Chromosomes, 69, 173
Chuño, 136
Cicer arientinum, 124
Citrullus, 180
Citrus, 184
Climatic change, 15–16, 212–213
Cocoa, 190
Coconut, 151, 159–167; crab, 166–167; pearls, 167
Cocos nucifera, 159
Coffea arabica, 189
Coffea canephora, 189
Coffee, 188–189
Cohen, Mark, 16–17
Coir, 161, 164
Coix lacryma-jobi, 108
Cola: acuminata, 191; *nitida,* 191
Colchicine, 204
Cole crops, 178
Colocasia esculenta, 149
Compositae, 171
Cook, James, 151
Copra, 162
Coprolites, 3
Corm, 149, 154
Corn, 89–105; blight, 102, 205; broom, 106; hybrid, 98–102; kinds of, 85; origin of, 95–98; uses of, 89, 102–103
Cortisone, 148
Cotton, 172–176
Cowpea, 124
Crab grass, 67
Crick, Francis, 207
Cruciferae, 178
Cucumber, 180
Cucumis, 180
Cucurbita, 178
Cucurbitaceae, 178

Darwin, Charles, 17, 99, 166
Dasheen, 148
Date, 29, 159
Dating, radiocarbon, 3
De Candolle, Alfonse, 200

Derris, 123
Dioscorea, 146, 148; *bulbifera*, 146
Dog, 8, 23, 37–39
Domestication, 197–200. *See also* animals
Donkey, 52
Durum, 70, 73

East, Edmund Murray, 99
Egg, 30
Eggplant, 181–182
Ehrlich, Anne, 206
Ehrlich, Paul, 206
Einkorn, 69–70, 72
Emmer, 70, 72
Energy, 211
Equus caballus, 51
Erosion, 211–212
Ethanol, 114
Euchlaena mexicana, 96
Euphorbiaceae, 143
Evolution, 197–200

Fagopyrum esculentum, 108
Famine, Irish potato, 137
Fats, 29, 30
Favism, 122
Fertility rites, 18–22
Fertilizers, 211
"First fruits" theory, 24–25
Fishermen, 15
Flannery, Kent, 14, 18
Food: producers, 1; production, 209
Food and Agriculture Organization, 207
Forests, 209
Fruits, 184–187; multiple, 151; pome and stone, 184

Gallus gallus, 55
Garbanzo, 124
Garlic, 183
Gasohol, 114–115, 211
Genetics, 197; engineering, 205; erosion, 205–206
Gerard, John, 90–91
Gibberella fujikuroi, 87
Gibberellin, 87
Glucose, 29
Gluten, 68
Glycine max, 129
Goats, 41–43
Gossopol, 176
Gossypium: arboreum, 173, 174; *barbadense*, 173, 174
Gourd, bottle, 180–181
Gramineae, 61

Grapefruit, 184
Grapes, 192–194
Grasses, 61–108; goat, 70, 204; as ornamentals, 67; for soil conservation, 67
Green Revolution, 77–78, 88, 205
Greenhouse effect, 209, 212
Gros Michel banana, 156
Groundnut, 127
Guanaco, 58
Guinea pig, 58
Gumbo, 173
Gymnosperms, 61

Hahn, Eduard, 23
Haploids, 204
Harlan, Jack, 68
Harris, Marvin, 46
Hazelnut, 187, 188
Helianthus: annuus, 169; *tuberosus*, 171
Hemp, Manila, 158
Hesperidium, 185
Heterosis, 98
Heyerdahl, Thor, 141
Hibiscus, 173
Honey, 29
Hordeum vulgare, 106
Horses, 51–53
Huaca Prieta, 9
Hunger, 1, 13, 208–209
Hunter-gatherers, 1
Hybridization, 98, 203
Hybrid vigor, 98

Ilex paraguariensis, 192
Iltis, H. H., 95n, 97n
Inbreds, 99–102, 205
Incaparina, 174–175
Inga, 120
Insecticides, 123, 211
International Board for Genetic Resources, 206
International Potato Center, 139
International Rice Research Institute, 87
Ipomoea: batatas, 139; *trifida*, 141
Irrigation, 12, 212
Isaac, Erich, 18
Isolation, 201

Jackfruit, 152–153
Jarmo, Iraq, 6
Jerusalem artichoke, 171
Jicama, 120
Job's tears, 108
Jojoba, 168
Jones, D. F., 99

Kale, 178
Keith, Minor Cooper, 155
Kellogg, John Harvey, 129
Kiwi, 187
Kohlrabi, 178
Kon Tiki, 141
Kumiss, 51

Labiatae, 195
Lactose, 29, 45
Lactuca sativa, 171
Lagenaria siceraria, 180
Lama: glama, 38; *pacos*, 58
Land, for cultivation, 212
"Last sheaf" theory, 24–25
Lathyrism, 122
Lathyrus sativus, 122
Legumes, 117–133
Lentils, 123
Lettuce, 171
Leucaena, 122
Liliaceae, 183
Lind, John, 185
Linnaeus, Carolus, 69, 90, 155
Lipids, 29–30
Llamas, 58–60
Locoweed, 123
Lonchocarpus, 123
Love apple, 181
Luau, 149
Lucerne, 121
Lupinus mutabilis, 123
Lycopersicon esculentum, 181
Lysine, 32

Macadamia, 188
Mace, 196
MacNeish, R. S., 8
Maize, 89–105. *See also* corn
Malnutrition, 28, 208
Malt, 108
Malvaceae, 173
Mangelsdorf, Paul, 96–97
Mangifera indica, 185
Mango, 185–186
Mani, 127
Manihot esculenta, 143
Manioc, 143–146
Marijuana, 27n, 61n
Maté, 191–192
May Day, 20
Meat, 30; synthetic, 133. *See also* animals
Medicago sativa, 121
Meleagris gallopavo, 58
Melons, 180
Mendel, Gregor, 99, 203, 207

Mexico, 8, 9–10
Michener, James, 14
Milk, 43, 45, 47–48, 50, 51
Millets, 108
Milo, 105
Mimosa, 121
Mint family, 195
Molecular biology, 205
Moraceae, 151
Morning glory family, 142
Mother goddess, 18–19
Mouflons, 39
Mulberry family, 151
Mule, 52
Muller, H. J., 203
Musa: acuminata, 155; *balbisiana*, 155; *paradisiaca*, 155; *sapientum*, 155; *textilis*, 158
Muscovy duck, 58
Mustard, 178
Mutation, 197, 203
Myristica fragrans, 176

Napoleon, 116
Near East, 6
Neolithic Revolution, 15
Niacin, 103
Nicotiana tabacum, 183
Nightshade family, 136, 181–183
Nishiyama, Ichizo, 141
Nitrogen fixation, 86, 118–119
Nodules, 119
Nutmeg, 196
Nutrition: defined, 29; human, 27–33
Nuts, 187–188

Oats, 106–107
Oceans, food from, 209
Oils, 29, 30, 103, 133, 162–163, 168–176, 178, 191
Okra, 173
Olea europaea, 169
Olive, 169
Onion, 183
Opaque-2 mutant, 104
Orange, 184
Orchid family, 195
Organic farming, 211
Oryza: glaberrima, 80; *rufipogon*, 80; *sativa*, 80
Ovis: aries, 39; *canadensis*, 39; *orientalis*, 39
Ox, 45
Oxytropis, 123

Pachyrrhizus, 120
Paddock, Paul, 213

Paddock, William, 213
Palms, 159
Papain, 187
Papaya, 187
Parsley family, 183, 195
Parsnip, 183
Pasteur, Louis, 192
Pastinaca sativa, 183
Peanut butter, 129
Peanuts, 126–129
Peas: field, 123; garden, 123–124; grass, 122; Indian, 122; rosary, 122; sweet, 121
Pecan, 188
Pellagra, 103
Pepitas, 180
Pepper: black, 195–196; chili, 182, 199; red, 182; sweet, 182; white, 196
Persea americana, 183
Peru, 9, 10
Pesticides, 211
Petroselinum crispum, 183
Petunia hybrida, 183
Phallic symbols, 19
Phaseolus: acutifolius, 124; *coccineus,* 124; *lunatus,* 124; *vulgaris,* 124–126
Philodendron, 149
Phoenix dactilifera, 159
Pig, 53–55
Pigweed, 108
Pincus, Gregory, 148
Pineapple, 186–187
Piper nigrum, 195
Pistachio, 188
Pisum sativum, 123
Plantain, 154
Poi, 149
Poinsettia, 143
Pollen, fossil, 16
Polyploidy, 69, 204
Population, 208–209
Potato: aerial, 146; Irish, 134–139; sweet, 139–142, 146; white, 134–139; yam, 146
Propagation, 10, 12, 137, 198; seed, 10; vegetative, 12
Prunus, 184
Psophocarpus tetragonolobus, 120
Pulse, 117, 118
Pumpkin, 180
Purseglove, John, 128

Quinua, 110

Radish, 178
Rape, 178
Raphanus sativus, 178

Recombination, 197, 201
Reed, Charles A., 16
Religion, early, 20–23
Reproduction, human, 20, 25–26; control of, 148
Research, 207
Resource depletions, 209–212
Rhizobium, 119
Rice, 79–88; paper, 87; wild, 88–89
Rick, Charles, 206
Rindos, David, 17
Rockefeller Foundation, 77
Rosaceae, 184
Rose family, 184
Rotenone, 123
Rubber, Pará, 143
Rum, 114
Ruminants, 36
Rutabaga, 178
Rutaceae, 184
Rye, 79, 106

Saccharum officinarum, 111
Sacrifice, 21–22
Safflower, 171
St. John's bread, 120
Sake, 86
Sauer, Carl O., 12, 15, 82
Scurvy, 185
Secale, 79, 106
Seed: banks, 206; wars, 206. *See also* propagation
Selection, 17, 197; conscious, 198–199, 203; unconscious, 18–19, 198
Serotonin, 156
Sex, in plants, 26
Sheep, 39–41
Shull, George Harrison, 99
Simmondsia chinensis, 168
Slavery, 111
Solanaceae, 181–183
Solanine, 139
Solanum tuberosum, 134, 139
Sorghum: bicolor, 105–106; *sudanense,* 106
Source, The, 14
Soybean, 129–133
Species, defined, 46n
Spelt, 73
Sperm, 26
Spices, 178, 182, 194–196
Spinach, 116
Spinacia oleracea, 116
Spurge family, 143
Squash, 180
Starch, 29

Sucrose, 29, 116, 164
Sugar, 111–116; beet, 115–116; cane, 111–115; coconut, 164; corn, 103; maple, 111
Sunflower, 169–172
Sus scrofa, 53
Sweet potato, 139–142, 146

Taboos, animal, 33, 45–46, 51, 54
Tamaulipas, 8
Tapioca, 145
Taro, 148–151
Tarwi, 123
Tea, 189–190; Paraguay, 191
Tehuacan, 8, 96, 129
Teosinte, 95–98, 206
Tetrapanax papyriferus, 87
Thailand, 6
Thaumatin, 111
Theobroma cacao, 190
Tobacco, 183, 212
Tofu, 132
Tomato, 181, 206
Tortilla, 103
Trinandwan, Sa-korn, 65
Triticale, 79
Triticosecale, 79
Triticum: aestivum, 73; *monococcum*, 69–70; *timopheevii*, 70; *turgidum*, 70, 73
Tuber, 137
Turkey, 8, 58
Turnip, 178

Umbelliferae, 183, 195
Ungulate, 36
United Fruit Company, 155
Urials, 39

Valery banana, 156
Vanilla, 195

Variability, 201, 205
Vegans, 31
Vegetables, 177–183
Vegetarians, 31
Vicia faba, 122
Vicuña, 58
Vigna, 126; *sesquipedalis*, 124; *sinensis*, 124
Visser, Margaret, 103
Vitamin: A, 63; B_1, 86; B_{12}, 31–32; C, 63, 185
Vitis vinifera, 192–194

Walnut, 188
Watermelon, 180, 204
Watson, James D., 207
Weatherwax, Paul, 89–90
Weeds, 67
Wheat, 67–78; archaeological, 72–73; changes in, 76; origin of, 69–72; rust, 77
Wine, 192–194
Wolf, 37
Wool, 40, 41, 43
Wright, H. E., Jr., 15–16

Xanthosoma, 149
X-rays, 203

Yam, 146–148
Yautia, 149
Yeast, 68
Yuca, 143
Yudkin, John, 27

Zanahoria, 183
Zea mays, 90
Zebu, 46
Zizania palustris, 88